Statistics and Econometrics for Finance

Series Editors
David Ruppert
Jianqing Fan
Eric Renault
Eric Zivot

More information about this series at http://www.springer.com/series/10377

Abdulkader Aljandali

Quantitative Analysis and IBM® SPSS® Statistics

A Guide for Business and Finance

 Springer

Abdulkader Aljandali
Accounting, Finance and Economics Department
Regent's University
London, UK

ISSN 2199-093X   ISSN 2199-0948   (electronic)
Statistics and Econometrics for Finance
ISBN 978-3-319-45527-3   ISBN 978-3-319-45528-0   (eBook)
DOI 10.1007/978-3-319-45528-0

Library of Congress Control Number: 2016953007

This Springer imprint is published by Springer Nature
The registered company is Springer International Publishing AG
The registered company address is: Gewerbestrasse 11, 6330 Cham, Switzerland

To Beybars

Preface

IBM SPSS Statistics is an integrated family of products that addresses the entire analytical process, from planning to data collection to analysis, reporting and deployment. It offers a powerful set of statistical and information analysis systems that runs on a wide variety of personal computers. As such, IBM SPSS (previously known as SPSS) is extensively used in industry, commerce, banking and local and national government education. Just a small subset of users of the package in the UK includes the major clearing banks, the BBC, British Gas, British Airways, British Telecom, Eurotunnel, GSK, TfL, the NHS, BAE Systems, Shell, Unilever and WHS.

In fact, all UK universities and the vast majority of universities worldwide use IBM SPSS Statistics for teaching and research. It is certainly an advantage for a student in the UK to have knowledge of the package since it obviates the need for an employer to provide in-house training. There is no text at present that is specifically aimed at the undergraduate market in Business Studies and associated subjects such as Finance, Marketing and Economics. Such subjects tend to have the largest numbers of enrolled students in many institutions, particularly in the former polytechnic sector. The author is not going to adopt an explicitly mathematical approach, but rather will stress the applicability of various statistical techniques to various problem-solving scenarios.

IBM SPSS Statistics offers all the benefits of the Windows environment as analysts can have many windows of different types open at once, enabling simultaneous working with raw data and results. Further, users may learn the logic of the programme by choosing an analysis rather than having to learn the IBM SPSS command language. The last thing wanted by students new to statistical methodology is simultaneously to have to learn a command language. There are many varieties of tabular output available, and the user may customise output using IBM SPSS script.

This guide aims to provide a gentle introduction to the IBM SPSS Statistics software for both students and professionals starting out with the package, although it is recognized that the latter group would probably be familiar with the content presented here. A second more advanced text building on this material will be beneficial to professionals working in the areas of practical business forecasting or market research data analysis. This text would doubtlessly be more sympathetic to the readership than the manuals supplied by IBM SPSS Inc.

London, UK   Abdulkader Mahmoud Aljandali

Introduction

This is the first part of a two-part guide to the IBM SPSS Statistics computer package for Business, Finance and Marketing students. This, the first part of the guide, introduces data entry, along with elementary statistical and graphical methods for summarizing data. The rudiments of hypothesis testing and business forecasting are also included. The second part of the guide presents multivariate statistical methods, more advanced forecasting and multivariate methods. Although the emphasis is on applications of IBM SPSS Statistics software, there is a need for the user to be aware of the statistical assumptions and rationale that underpin correct and meaningful application of the techniques that are available in the package. Therefore, such assumptions are discussed and methods of assessing their validity are described. Also presented is the logic underlying the computation of the more commonly used test statistics in the area of hypothesis testing. However, the mathematical background is kept to a minimum.

This, the first part of the IBM SPSS Statistics guide, is itself divided into five sections. Throughout, real and manually contrived data sets are used which could be accessible via the publisher's website. Part I introduces IBM SPSS Statistics. A data file is created and saved. Different levels of data measurement are discussed, in that the selection of appropriate analytical tools is dependent upon them. Elementary descriptive statistics are computed, and the user is introduced to the graphics facilities available in IBM SPSS Statistics. Much can be achieved in a short while, once the user is familiar with the individual windows and files of the software.

A lot of information can be gleaned about the characteristics of collected data by graphical means, for example, many statistical routines require data to be normally distributed. The first chapter of Part II expands on the graphics facilities in IBM SPSS Statistics. Similarly, frequency tables and cross-tabulations of variables assist in detecting data characteristics, and these are the subject matter of Chap. 3. Chapter 4 discusses the coding of data entry into a computer package. In many data-gathering exercises, there are missing values. IBM SPSS Statistics offers a very simple procedure for declaring missing values and, more generally, for labelling individual variables and their values. Sometimes, variables have to be transformed into other variables, e.g. the conversion of one currency into another. These features of IBM SPSS Statistics conclude Part II.

Part III introduces and describes hypothesis tests. After a review of hypothesis testing, major parametric (Chap. 5) and nonparametric methods (Chap. 6) are described and illustrated by application. Parametric methods make more rigid assumptions about the distributional form of the gathered data than do nonparametric methods. However, it must be recognised that parametric methods are more powerful when the assumptions underlying them are met.

Part IV introduces elementary forecasting methods. Two-variable regression and correlation are illustrated in Chap. 7, and the assumptions underlying the regression method are stressed. Many of these assumptions may be assessed graphically by any methods previously described in Part II. Chapter 8 describes and illustrates two methods of time series analysis—seasonal decomposition and one-parameter exponential smoothing. The practical utility of both time series methods is discussed.

Part V comprises a chapter that presents other features of IBM SPSS Statistics that are likely to be useful, once the user is familiar with the basics of the package. The user is encouraged to access the IBM SPSS Statistics Help system. This part also introduces primary and secondary data in addition to various sources that a student in Business, Finance or Marketing course might need as part of their curriculum learning.

Once users are familiar with the methods described in this text, the assumptions that underpin them and the windows that access the routines, then they may fruitfully experiment and often learn on their own. For example, it would take a guide far larger than this just to describe all of the graphics capabilities of the package and associated styles of presentation. This guide provides a sufficient depth of introduction for users of the package to investigate alternative graphical forms. Indeed, the purpose of this guide is generally to provide sufficient statistical background for the user to be able to perform meaningful analysis, to enable the user to gather an insight about the characteristics of gathered data and to encourage him/her to experiment with allied features of the IBM SPSS Statistics system.

Acknowledgements

Welcome to the first edition of *Quantitative Analysis and IBM® SPSS® Statistics*.

I would like to take the opportunity to thank the many people who have contributed to this book. Professor John Trevor Coshall takes full credit for his support in the writing of the first edition of this manuscript. The current textbook is inspired by the many SPSS handouts that John wrote for a variety of courses. John's objective was always to enable students of Business, Finance and Marketing to actively engage in the quantitative analysis discipline by undertaking their own research. Reading about various statistical assumptions and techniques can be interesting, but the core learning would be to use those same techniques and make sense of them, and John was restless in achieving the latter.

I would also like to thank my colleague, Ibrahim Ganiyu, for initiating the idea of book writing in a subject that falls within my area of expertise. His experience in terms of book publishing set me en route to write the current manuscript. Ibrahim's support was second to none and his insights immensely helpful in ensuring a strong foundation of the book editing process. We would both agree that we owe it to students to produce learning materials that are accessible and relevant.

I am indebted to my coach Alex Lawson; without his help I wouldn't have found the mental strength and balance to carry out such an immense task. Alex has been a much needed sounding board, and that has made a significant difference, especially when things didn't go to plan.

Finally, I would like to thank the team at Springer USA for their continuous support. In particular, I would like to acknowledge Mike Penn and Rebekah McClure who have worked closely with me to produce this edition. Thank you all.

Contents

List of Figures

List of Tables

Part I
Introduction to IBM SPSS Statistics

Chapter 1
Getting Started

The objective of this first chapter is to introduce some of the basic features of IBM SPSS Statistics. Essentially, much can be achieved in a short space of time once the user has become used to accessing and making selections from the various descriptive menus and dialogue boxes that are available. Most tasks may be performed by simply pointing and clicking the mouse.

In this chapter, a small data file is to be created in IBM SPSS Statistics and saved on memory stick or hard drive. The data involve the population sizes and number of retail shops in various European countries. There is a general description of basic statistics such as the mean and standard deviation, which are then computed for the above variables. The charting facility in IBM SPSS Statistics is introduced and a plot of the number of shops against the countries' population sizes is generated.

1.1 Creation of an IBM SPSS Statistics Data File

IBM SPSS Statistics can read data input files from a variety of external sources such as Excel and SPSS data files created on other operating systems. However, in this section, we are going to create and save our own data file. The IBM SPSS Statistics Data Editor permits the entry of data and the creation of a data file. The Data Editor is a simple spreadsheet-like facility that opens automatically when you start an IBM SPSS Statistics session. However, please note that the Data Editor does not operate like a spreadsheet, for example, you cannot enter formulae into it. Table 1.1 presents the data which will be the input of our IBM SPSS Statistics data file.

The population sizes and number of retail outlets in Table 1.1 are called *numeric* variables. Valid numeric values include numerals, a decimal point and a leading plus or minus sign. The maximum width for numeric variables in IBM SPSS Statistics is 40 characters and the maximum number of decimal places is 16. *The names of the nine countries* in Table 1.1 are called *string* or *alphanumeric* variables. Valid string values involve letters, numerals and some other characters. String variables with

© Springer International Publishing Switzerland 2016
A. Aljandali, *Quantitative Analysis and IBM® SPSS® Statistics*,
Statistics and Econometrics for Finance, DOI 10.1007/978-3-319-45528-0_1

Table 1.1 Populations and number of retail outlets in selected countries (year 2015)

Name of country	Population size (000's)	No. of retail outlets
Belgium	11,292	69,682
Denmark	5660	21,745
Finland	5474	23,374
France	64,216	318,998
Germany	80,948	286,060
The Netherlands	16,902	100,270
Norway	5167	33,711
Sweden	9731	42,434
United Kingdom	64,708	279,726

eight or fewer characters are called short strings; those with a width of more than eight characters are long strings.

We shall need to name the three variables - *name of country, population size and number of retail outlets* in IBM SPSS Statistics. Variable names must begin with a letter and be unique. Blanks and characters such as *, !, ' and ? may not be used. However, certain other characters are permitted, for example, STORE#1 and OVER$200 are legitimate variable names. Variable names are not case sensitive, so OLDVAR, oldvar and OldVar are the same in IBM SPSS Statistics.

The names chosen for the three variables of Table 1.1 and which will be used in our data file are shown below in capital letters:

- CTRY—name of country
- POPN—population size
- RETAIL—no. of retail outlets

As shown in this section, it is possible in IBM SPSS Statistics to attach more meaningful labels to these variable names and which will be reported on the generated output. For example, we may wish the variable name POPN to have the label POPULATION SIZE attached to it in our statistical output.

1.1.1   The IBM SPSS Statistics Data Editor

Upon entry to IBM SPSS Statistics, you will be presented with the *Data Editor Window* which contains the menu bar:

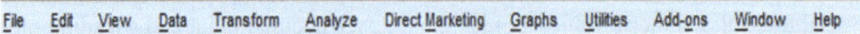

Amongst other things, the above menu bar is used to open previously created files, create new files (as we wish to do here), produce charts, choose statistical routines and select other features of the IBM SPSS system. Items can be selected from the menu bar via the mouse.

Note that:

- The rows of the Data Editor window are cases.
- The columns represent the study variables.
- Cells may only contain data values (numeric or string).
- Formulae are not permitted.

In the present example, the rows will be each of the nine countries of Table 1.1. The columns will refer to the variable names CTRY, POPN and RETAIL. We are going to use the Data Editor to enter the variable names, label these names and enter the raw data of Table 1.1. A blank Data Editor is shown in Fig. 1.1. In the bottom left hand corner of the Data Editor, click the 'Variable View' tab, which gives rise to the dialogue box of Fig. 1.2.

The name of the first variable is CTRY, so enter this into the first row of the Variable View in the column labelled Name. Via the Enter key, the dialogue box of Fig. 1.3 is now generated. By default, IBM SPSS Statistics assumes that variables are numeric. The width of 8 refers to the maximum number of characters to be used, including one position for any decimal point. The numeral 2 refers to the number of decimal positions for display purposes and appears in the Decimals column of Fig. 1.2. The variable CTRY is, however, a string variable. Click the small grey box next to the word numeric in Fig. 1.3 which now produces the Variable Type dialogue box of Fig. 1.4. In this latter dialogue box, click the option String and then the OK button. This alters the variable type for CTRY as shown in Fig. 1.5.

It should be noted that the user may start off by typing data straight into the Data Editor of Fig. 1.1, without first defining the variable names. In this case,

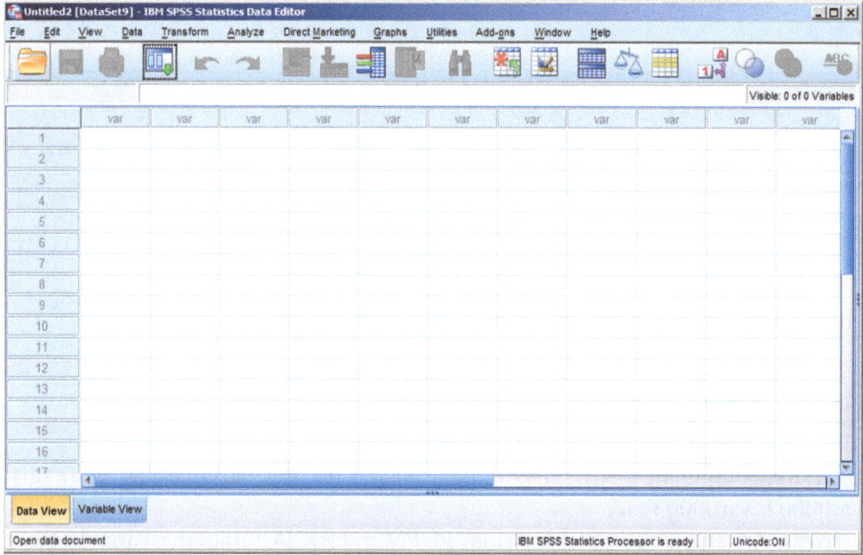

Fig. 1.1 The IBM SPSS Statistics Data Editor

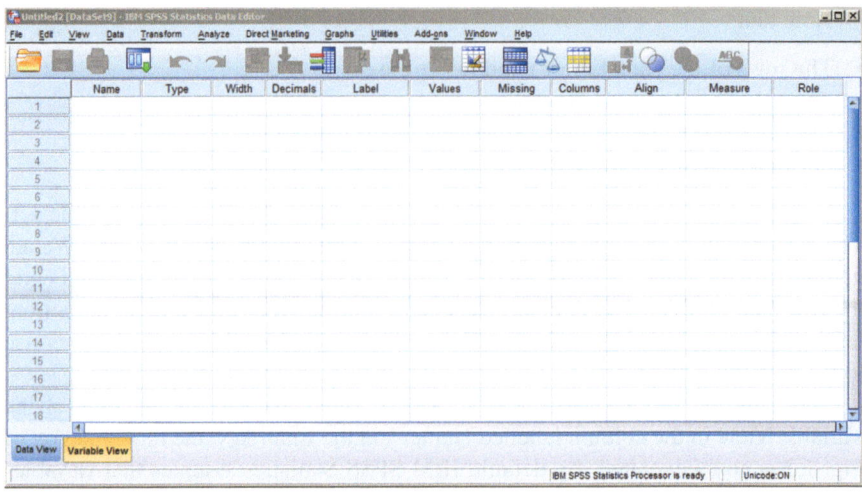

Fig. 1.2 The IBM SPSS Statistics Variable View

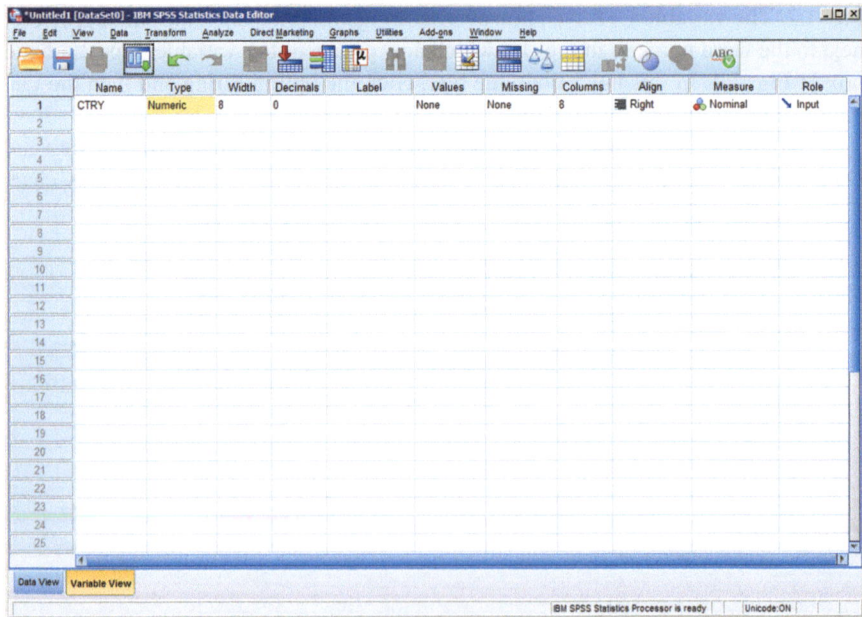

Fig. 1.3 Defining a Variable

IBM SPSS Statistics will give default names to the variables as var00001, var00002, var00003 etc.

Next, one enters the variable names POPN and RETAIL into the Variable View. Both of these variables are numeric. If we chose the number of decimal places as 2,

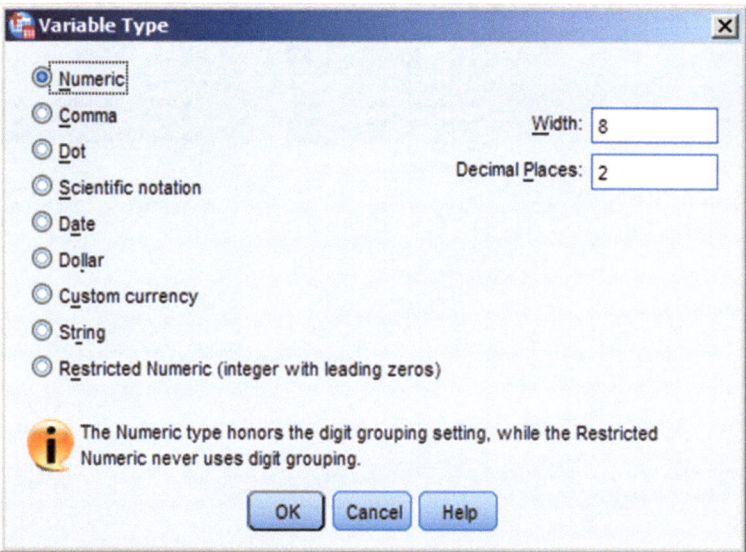

Fig. 1.4 The Variable Type Dialogue Box

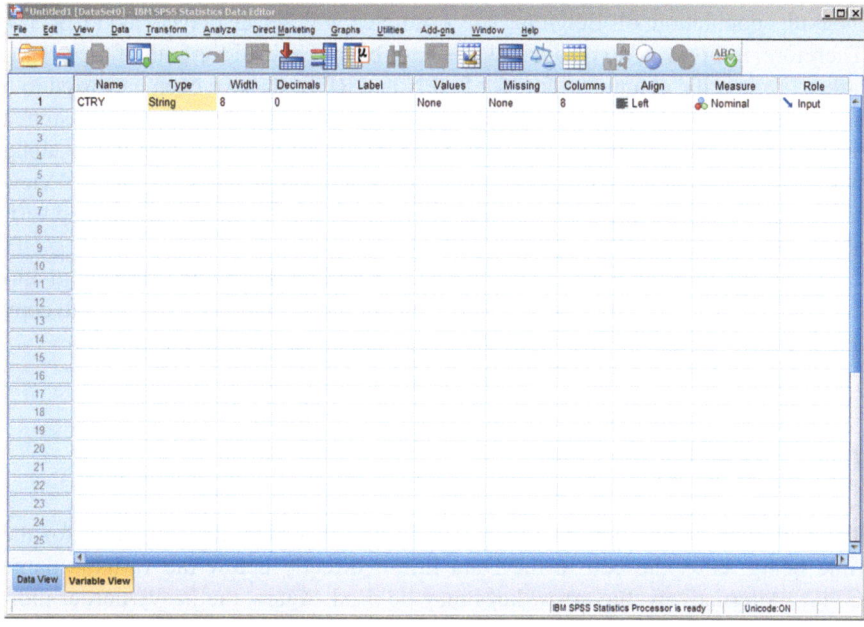

Fig. 1.5 Defining a String Variable

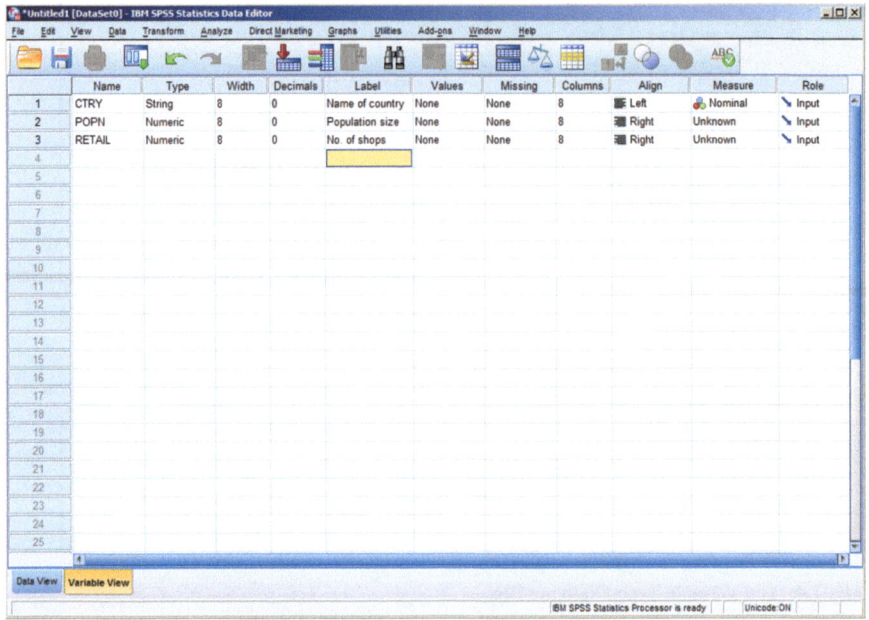

Fig. 1.6 Defining Numeric Variables

then the population of Belgium, for example, will be displayed as 11292.00.
Therefore, in Fig. 1.6, no decimal places have been specified for both of these variables. Further, the column widths for POPN and RETAIL have been narrowed to 5
and 6 respectively. In the column titled Label, all three variables have been assigned
labels which will appear on any IBM SPSS Statistics output. These labels along
with the variable names will appear on the generated output. Clicking the Data View
tab returns the user to the Data Editor as shown in Fig. 1.7, wherein the defined variable names appear.

A final point is that it is possible to copy the attributes from one variable to others. Simply click the cell in the Variable View for the attribute that you want to copy
and use the copy and paste options that are found under the Edit menu item.

1.1.2 Entering the Data

The data may be entered in virtually any order. However, for simplicity for the
time being, click the cell in the Data Editor directly below the variable name
CTRY. Alternatively, the arrow keys may be used. Again, the heavy border indicates that the cell is active. The variable name and the row number appear in the
upper left hand corner of the Data Editor.

From Table 1.1, type in Belgium into cell 1: CTRY and press the Enter key. The
data value now appears in that cell and cell 2: CTRY becomes active, awaiting a data

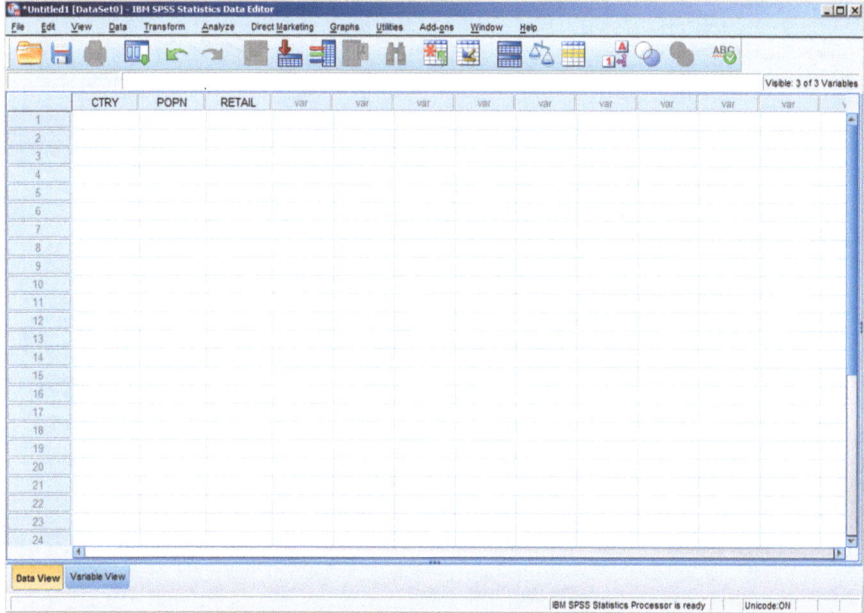

Fig. 1.7 The IBM SPSS Statistics Data Editor with variable names defined

value entry. It should be noted that after entering the value for one variable for a particular case, the cells of the other variables for that case become "system missing", as
indicated by the full stop in those cells. These latter cells are simply awaiting data
entry. Having entered all the values for the variable CTRY, click the top cell for the
variable POPN (or use the arrow keys to arrive at this cell location) to start entering
values for this variable. Continue entering the data values for the three variables.

1.1.3   Saving the Data File

Any changes made to a data file in the Data Editor window last only for the duration
of your IBM SPSS Statistics session or until another data file is opened. Having
fully defined our file, we now wish to save it. From the Data Editor click:

File
    Save A…

a dialogue box will now appear with the title 'Save Data As' and which is shown in
Fig. 1.8. Suppose *the file* on which the data are to be saved is in the E: drive. We
need to change to this drive. This is achieved by selecting the appropriate alternative
in the box labelled 'Look in'.

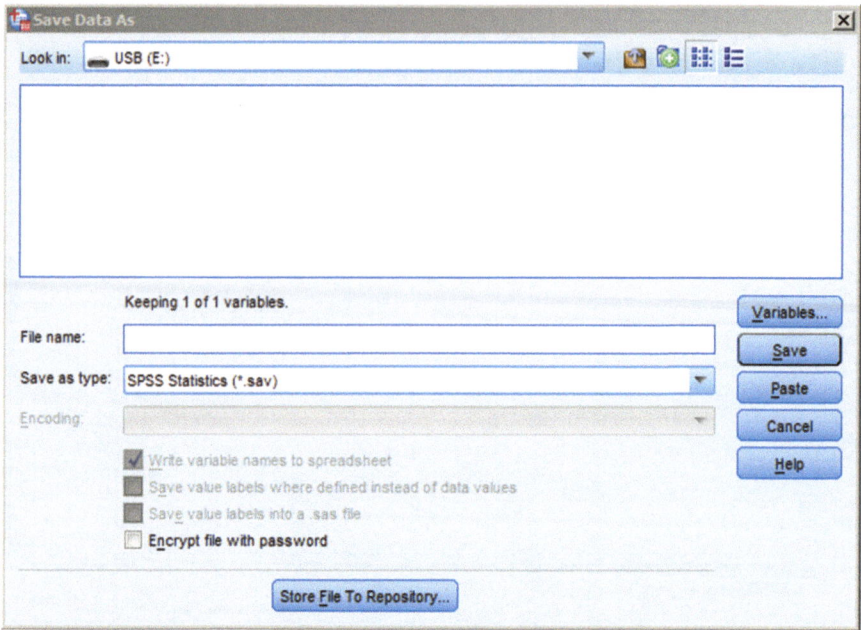

Fig. 1.8 The IBM SPSS Statistics Window for Saving Data

Data files created and/or saved in IBM SPSS Statistics have the extension .
SAV. We need to name our data file - say RETAIL.SAV. Enter this in the File Name
box and click OK. The data file is now saved on the E: drive with the name RETAIL.
SAV. Note that all variable labels etc. are also saved. It is always wise to save data
every quarter of an hour or so, in case of misfortunes such as a computer crash or a
power cut. On future occasions, click:

File
   Save

because the system will now know that the data file is to be saved on the E: drive.
Only if the drive is to be changed click:

File
   Save As...

Should you ever forget to save any type of IBM SPSS Statistics file, you will be
prompted to do so on leaving IBM SPSS Statistics for Windows.

1.2 Descriptive Statistics

However complex the statistical routines that are to be employed during data analysis, it is always prudent to perform an initial examination of the raw data. Such an
examination might highlight data input errors or the failure to note missing values,

which is always a possibility in the coding of the results of large surveys. Some statistical methods in IBM SPSS Statistics assume that the sample data are taken from a population that is normally distributed. Computation of some of the descriptive statistics described in the next sub-sections, along with some of the graphical procedures introduced in the next chapter allow assessment of this assumption.

1.2.1   Some Commonly Used Descriptive Statistics

Data may be characterized by two useful types of measure. Firstly, *measures of central tendency* (sometimes also called averages or measures of location) attempt to locate a typical value about which the data cluster. Secondly, there are measures indicative of how spread out or scattered a data set is. The latter are called *measures of dispersion*. Both types of measure are numerical quantities compatible with the data and are measured in the same units as the data themselves.

The most widely used and familiar measure of central tendency is the *arithmetic mean*, commonly referred to as simply the *mean*. Most commercial and business data are sampled data drawn by some method from an underlying population, which is too costly, large or time consuming to access. The notation $\bar{x}$ is commonly used to denote the sample mean and the notation μ (the Greek letter 'mu') is commonly used to denote the population mean. A typical problem is that given a value for $\bar{x}$, what inferences may be made about the population mean? For example, if a sample of $n=1000$ households in a borough was found to spend a mean of $\bar{x} = £300$ per year on domestic insurance, what may be inferred about the population mean expenditure on domestic insurance in the borough? Such problems are discussed in later chapters.

Suppose we have a sample of n observations. Denoting the first reading as x_1, the second reading as x_2 etc., then the sample mean based on n observations is defined as:

$$\bar{x} = \sum_{i=1}^{n} x_i.$$

In general, the arithmetic mean is the sum of the observations divided by the number of observations. For example, if a sample of $n=7$ observations yielded the following annual expenditures on domestic insurance:

$$295 \quad 300 \quad 304 \quad 302 \quad 355 \quad 256 \quad 302 (£\text{'s})$$

then the sample mean is $2114/7 = £302$. Especially in the case of small samples, the mean can be influenced by extreme values. For example, if the weekly salaries of five interns were:

$$334 \quad 330 \quad 340 \quad 350 \quad 670 (£\text{'s})$$

then the sample mean may be computed as £404.80. Four of the wages are below the mean while that of the fifth intern is well above it. The mean is not really representing the data adequately. The median is a measure of central tendency that is ideally suited to this latter situation. The median is defined as the middle reading when the data set is arranged in size order. For example, when ordered from low to high, the seven annual expenditures on domestic insurance become:

$$256 \quad 295 \quad 300 \quad 302 \quad 302 \quad 304 \quad 355(£'s).$$

The median is thus the fourth reading of £302. Obviously the same answer would be obtained if the data were arranged from high to low. Note that the median of the five weekly salaries previously reported is £340 and is more reasonable as an average than the mean of £404.80. If the data consists of an even number of readings, then no unique middle value exists. In this situation, IBM SPSS Statistics adopts the convention of defining the median as the mean of the middle two observations.

Another measure of central tendency that may be mentioned is the mode. The mode is defined as the reading that occurs with the greatest frequency or most often. The sample on insurance expenditures is small for the purposes of illustration, however the modal expenditure is £302 as this reading occurs twice (a frequency of two), while the other readings occur once. Of course, it is possible for a set of data not to possess a mode if all the observations are numerically unique.

Turning to measures of dispersion or spread, the simplest is the range which is the difference between the numerically largest and smallest observations in the gathered data. The range of our seven expenditures on domestic insurance is, therefore, £355-£256=£99. The most widely used measure of dispersion in Statistics is the standard deviation, which is based on the mean. The square of the standard deviation is called the variance. The notation s^2 is commonly used for the variance of sample data; the notation σ^2 (the Greek letter 'sigma' squared) being employed for the variance when population data are involved.

The sample variance is defined as:

$$s^2 = \frac{1}{n}\left[\sum_{i=1}^{n}(x_i - \bar{x})^2\right]$$

where again, $\bar{x}$ is the sample mean and n is the number of observations. The standard deviation is the square root of the above formula. The variance as defined above is thus the mean of the squared deviations from $\bar{x}$. It might be noted that the sum of the deviations from mean, namely $\sum_{i=1}^{n}(x_i - \bar{x})$ is always equal to zero, so the latter expression is not useful in defining a measure of spread. This goes some way to explaining why the sum of the squared deviations rather than the sum of the actual deviations is used in the formula for the sample variance.

Returning to the insurance data, which have as a mean $\bar{x} = £302$:

x_i :	295	300	304	302	355	256	302
$(x_i - \bar{x})$:	−7	−2	2	0	53	−46	0
$(x_i - \bar{x})^2$:	49	4	4	0	2809	2116	0

we find that $\sum (x_i - \bar{x})^2 = 4982$, whereby the sample variance $s^2 = 4982/7 = 711.714$. Taking the square root, the sample standard deviation is $s = £26.68$.

The standard deviation is a measuring unit for spread in a given data set. In the above example, we may say that one standard deviation (1s here) equals £26.68. We can use this fact as a conversion factor to measure spread of the domestic insurance expenditures, not in £'s but rather in s units. It is just like knowledge of the pertinent exchange rate permits conversion of £ sterling into euros. The lowest reading in our sample is £256, which is £46 below the mean of £302. If 1s=£26.68, then £46 is worth (46/26.68)s=1.72s. We say that our sample data extend 1.72 standard deviations (1.72s) below the sample mean. Similarly, the highest reading in the sample is £355, which is £53 above the mean of £302. If 1s=£26.68, then £53 is worth (53/26.68) s=1.99s. Our sample data extend 1.99 standard deviations (1.99s) above the mean.

The standard deviation, s, as a measure of spread permits the comparison of spread or dispersion inherent in different samples. For example, the lengths of industrially manufactured plastic boxes may be measured in centimetres. The weights of these same boxes may be measured in grams. It is impossible to say that a spread of 4 cm in the lengths of the boxes is twice the spread of 2 g in their weights, since the units of measurement are different. However, if the spread of both the lengths and weights are converted to s units, then comparisons about spread or variability may be made. Another measure of dispersion is the inter-quartile range, which is often used in conjunction with the median. The inter-quartile range is discussed later along with an associated graphical representation called the boxplot. The appropriateness or otherwise of various summary statistics depends on the level of measurement of the data.

1.2.2   Levels of Measurement

A traditional classification of levels of measurement into four scales is attributable to Stevens (1946). These scales are:

The nominal scale: This is the most basic level of measurement and involves the classification of items into two or more groups that are as homogeneous as possible. For example, students might be classified according to the level of study (undergraduate, postgraduate etc.). When data are coded for input into a datum file, codes such as 1 and 2 might be applied to undergraduate and postgraduate studies respectively. These numerals are merely identifiers and no meaning can be attached to their numerical size. In market research surveys, the most common nominal responses occur to questions involving the possible responses "yes" (codes as 1, say), "no" (coded as 2) and "don't know" (coded as 3).

The ordinal scale: This involves ordering items according to the degree to which they possess a particular characteristic. For example, an attitude measurement scale could be applied to consumers who are unfavourable, neutral or favourable to accept a new style of product packaging. Codes of 1, 2 and 3 could be applied to these possible responses. We know that a code of 3 is more favourable than a code 1, but not three times more favourable. Also, the difference between codes of 1 and 2 is not assumed to be the same as the difference between codes of 2 and 3.

The interval scale: If it is possible to rank items according to the degree to which they possess a particular characteristic and the differences (or intervals) between any two numbers on the scale have meaning, we have stronger level of measurement than ordinal. If we know how large the intervals between all items are on the scale and such intervals have substantive meaning, we have achieved interval measurement. The unit of measurement and the zero point in interval measurement are arbitrary. Temperature scales such as Fahrenheit and Celsius are examples of interval measurement. When measuring temperature, the zero point and unit of measurement are arbitrary; they are different for the aforementioned two scales. Interval scales permit examination of the differences between items but not their proportionate magnitudes. For example, 30 C is not twice as hot as 15 C. Converting these two figures to Fahrenheit further illustrates this point; the first figure is no longer double the second.

The ratio scale: When we add a true zero point as the origin of an interval scale, we have a ratio scale. The ratio of any two scale points is independent of the unit of measurement used. If two objects are weighed in pounds and grams, the ratio of the two pound weights would equal the ratio of the 2 g weights.

As stated earlier, the level of measurement controls the descriptive statistics and statistical procedures that might be meaningfully applied to data. Table 1.2 summarizes statistical measures that are appropriate at various levels of measurement. For example, it would make little sense to use the mean as a measure of central tendency if the data are nominal. (In that nominal data are unordered, there can be no measure of central tendency; however, the mode may be an appropriate summary statistic).

At the ordinal level of measurement, the measure associated with nominal measurement may also be used. At the interval level of measurement, measures associated ordinal and nominal measurement may also be used. Some of the IBM SPSS Statistics Help menus, especially those associated with statistical hypothesis

Table 1.2 Statistical measures at various levels of measurement

Measures of:			
Measurement level	Central tendency	Spread	Correlation
Nominal	–	–	Contingency coefficient
Ordinal	Median	Inter-quartile range	Spearman's rank
Interval	Mean	Standard deviation	Pearson's r
Ratio	All the above	All the above	All the above

testing, as well as various dialogue boxes, use Stevens' classification in statements about the levels of measurement necessary for particular procedures to be used.

1.2.3 Descriptive Statistics in IBM SPSS Statistics

The Descriptives procedure in IBM SPSS Statistics computes univariate summary statistics (that is summaries for one variable at a time). From the menu bar in the Data Editor click:

Analyze
  Descriptive Statistics
    Descriptives...

this opens the Descriptives dialogue box shown in Fig. 1.9.

The numeric variables initially appear in the source list to the left. Select all the variables for which you require descriptive statistics. Use the mouse and click POPN. Click the right pointing arrow in this dialogue box and POPN now appears in the Variable(s) box as shown in the next page. Repeat the procedure for the variable RETAIL. Click the Options... button in the *Descriptives dialogue box* to select the summary statistics required. This invokes the *Descriptives: Options dialogue box* illustrated in Fig. 1.10. In the present example, Mean, Std. deviation, Minimum, Maximum, Kurtosis and Skewness were selected, by clicking the mouse in the appropriate squares. Mean, Minimum, Maximum and Std. deviation are the defaults, so crosses already appear in the selection boxes. Section 1.2.4 describes what is meant by skewness and kurtosis. Click the Continue button to return to the *Descriptives dialogue box* of Fig. 1.9. Statistical (and graphical) output is displayed in the *IBM SPSS Statistics Viewer*, which is shown in Fig. 1.11. It is possible to save

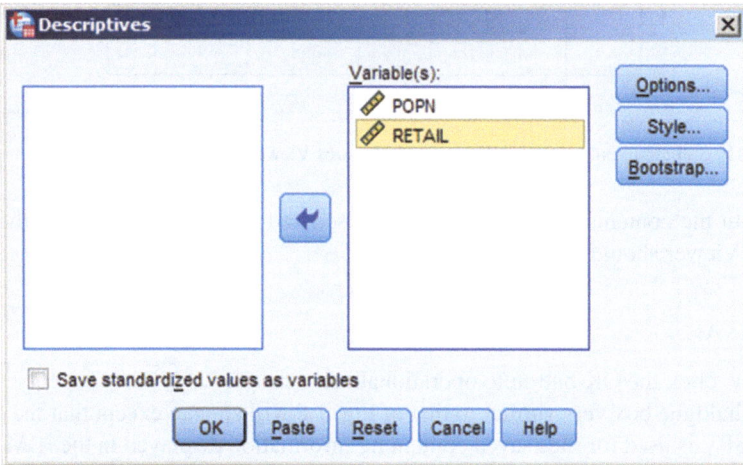

Fig. 1.9 The Descriptives dialogue box

Fig. 1.10 The
Descriptives: Options
dialogue box

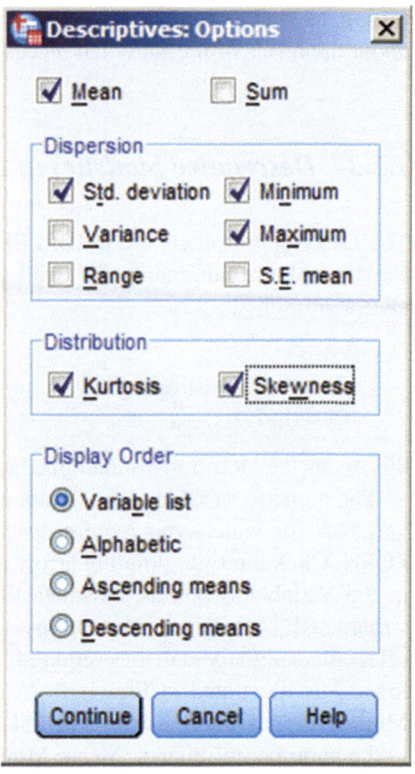

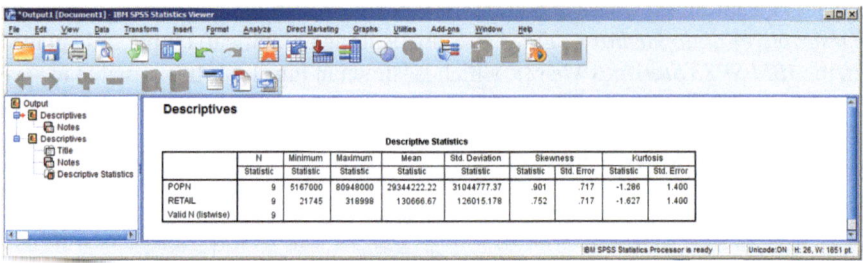

Fig. 1.11 Statistical output in the IBM SPSS Statistics Viewer

and edit the contents of the Viewer, as is discussed later. The contents of the IBM
SPSS Viewer should be saved via:

File
    Save As…

    Now click the OK button to operationalize.
    A dialogue box very similar to that of Fig. 1.8 will appear, except that the exten-
sion .SPV is used for files saved containing information displayed in the IBM SPSS
Statistics Viewer.

1.2.4   A Discussion of the Results

It should be noted that the variance quoted in the IBM SPSS Statistics Descriptives output is the unbiased estimator of the population variance, namely:

$$\text{Estimate of population variance} = \frac{ns^2}{n-1},$$

where s^2 is the sample variance, previously defined.

A distribution that is not symmetric is said to be *skewed*. If the longer tail is towards smaller values, the distribution is said to be *negatively skewed* and vice versa for *positive skew*. A perfectly symmetric distribution has a skewness of zero. A skewness of zero does not imply that the data are normally distributed, only that the distribution of data values is symmetric. A non-zero skewness, however, does suggest that the data are (to a relative extent) non-normal. Kurtosis refers to whether data tend to pile up around the centre of the distribution for a given standard deviation. If the data cases cluster around the central point less than is the case for the normal distribution i.e. the observed distribution is flatter, then the observed distribution is said to be *platykurtic* and the value of the kurtosis coefficient reported by IBM SPSS Statistics will be negative. If the data cases cluster more than is the case for the normal distribution i.e. the observed distribution is more peaked, then the observed distribution is said to be *leptokurtic* and the value of the kurtosis coefficient will be positive. In between these two extremes is the *mesokurtic* normal distribution. The kurtosis coefficient is zero in the mesokurtic case. If the data are normally distributed, the value of the kurtosis coefficient is 3.

Examination of Fig. 1.11 suggests that neither POPN nor RETAIL is regarded as normally distributed variables. The skewness of both variables is positive; indicative that the data is skewed to the right which means that the right tail of the distribution is long relative to the left tail. The same applies to the RETAIL variable. The kurtosis for RETAIL (-1.627) is smaller than the kurtosis for POPN (-1.286) which indicates that the distribution for the latter variable is less peaked (leptokurtic) than the normal distribution.

1.3   Creation of a Chart

It might be expected that countries with larger population sizes (POPN) would possess more shops (RETAIL) in order to meet potential demand. Is this the case and if so, do these variables increase in a linear fashion? A simple way of examining these contentions is to construct a chart of the RETAIL against POPN. We will produce a simple scatterplot of these two variables.

The Data Editor window must still be active. If it is not, for example, because you have logged off, call up RETAIL.SAV from the Data Editor via:

File
  Open
    Data…

scroll through the drives until you access the drive where RETAIL.SAV is located and select this file. Assuming that you have not terminated this IBM SPSS Statistics session, the Data Editor window is still active. Click:

Graphs
  Legacy Dialogs
    Scatter/Dot…

This generates the *Scatterplot dialogue box* of Fig. 1.12. In the simple scatterplot option each point represents the values of two variables. Click this option then click the Define button. This produces the *Simple Scatterplot dialogue box* of Fig. 1.13. Click the variable RETAIL and click the top arrow to define RETAIL as the Y Axis variable. Click POPN and click the second arrow to define POPN as the X Axis variable. In the dialogue box of Fig. 1.13, click the Titles… button to add a title the graph. You could Label Cases by the name of the CTRY, but this would clutter up the diagram somewhat. Upon clicking the OK button in the *Simple Scatterplot dialogue box*, the scatterplot is presented in the IBM SPSS Statistics Viewer (with the desired title), as shown in Fig. 1.14.

1.4 Basic Editing of a Chart and Saving it in a File

It is possible to change the characteristics of the plot in Fig. 1.14, for example, you may wish to change axis scaling, the colours used, the styles of shading, the position of titles etc. This is called the process of *editing* a chart which is performed in the *Chart Editor*. Double click inside the plot of Fig. 1.14 to access the Chart Editor.

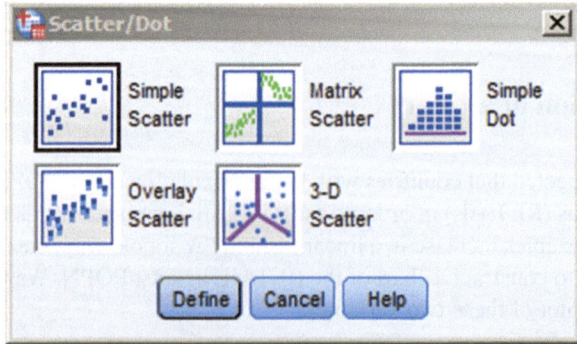

Fig. 1.12 The Scatter/Dot dialogue box

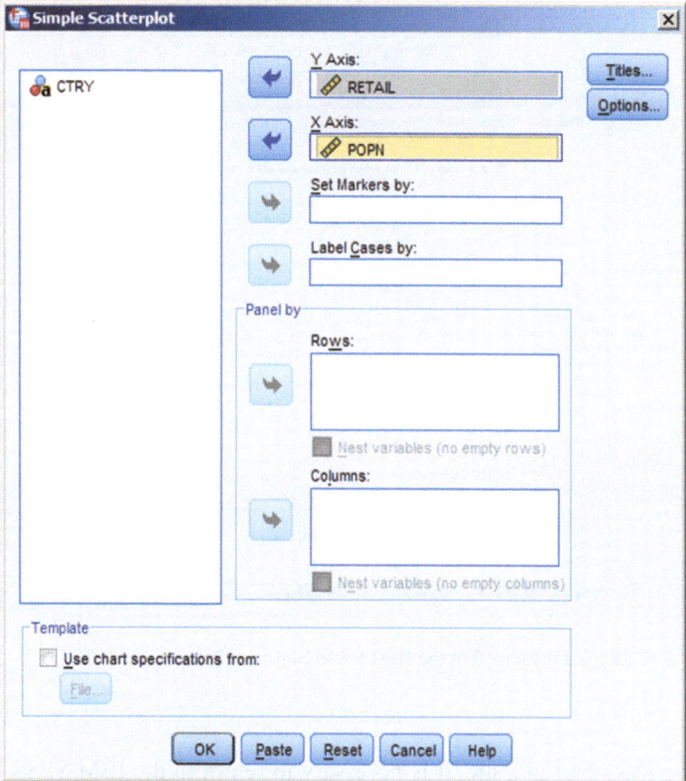

Fig. 1.13   The Simple Scatterplot dialogue box

Figure 1.15 presents the above scatterplot in the Chart Editor. Suppose we wish to change the circles on this plot to another format. The third icon from the left at the top of the Chart Editor ▦ is called the 'Show Properties Button'. Click any one of the circles shown on the above scatterplot. All circles become highlighted as indicated by the blue circle that surrounds them. Click the 'Show Properties Button' to generate the Properties dialogue box of Fig. 1.16. In this dialogue box, it is possible to change the symbol used in the scatterplot, via the Marker Tab. Click this tab to generate Fig. 1.17. Click the Type button to change the display from a circle to whatever you wish. You can fill in the new symbol that you have selected if you want by clicking the Fill box and choosing the colour black, say, from the palette. Similarly, the selected symbol may be resized via the Size options. To operationalize, click the Apply and Close buttons.

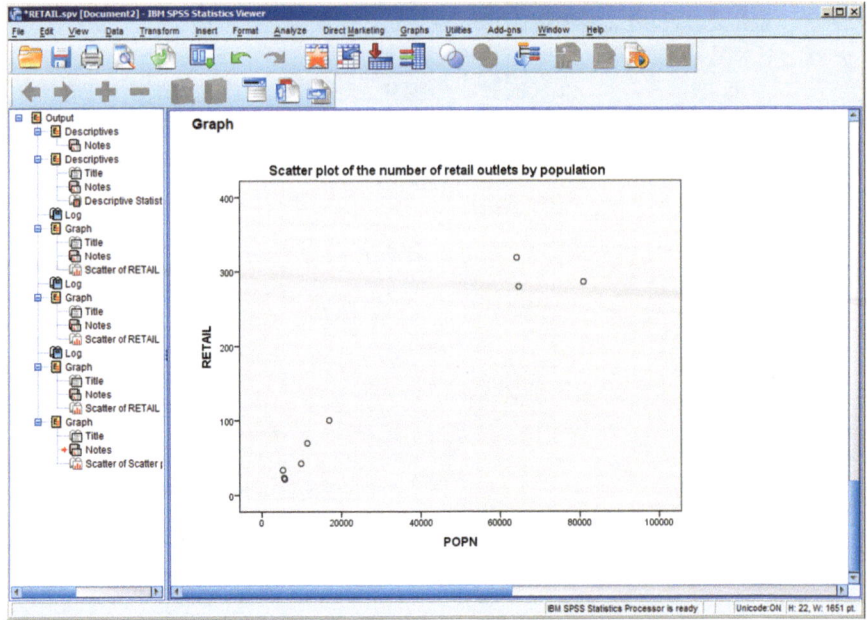

Fig. 1.14   A scatterplot presented in the IBM SPSS Statistics Viewer

To save this chart in a file, it is necessary to return to the IBM SPSS Statistics Viewer, by clicking the black cross in the top right hand corner of the screen. Once back at the Viewer, right click once inside the scatterplot and click:

Export...

from the options available which produces the Export Output dialogue box of Fig. 1.18.

The above dialogue box is split into two halves—the Document section enables the user to save output in, say, Word format; the Graphics section which is currently inactive, allows the user to save graphs/charts as separate files which may be imported into a word-processing package. In the Document section and under the heading Type, choose:

Select none (Graphics Only) which activates the Graphics half of the above dialogue box and results in Fig. 1.19.

Note that the default is to save the graphic in .JPEG format. This format may be changed via the Type option in the Graphics segment of the above dialogue box. The user can select the location for saving this graphics file via the Browse button. Here, the G: drive was selected. Click the OK button to operationalize. To insert this graphic into Microsoft Word, open that package and select the Insert tab, then Picture.

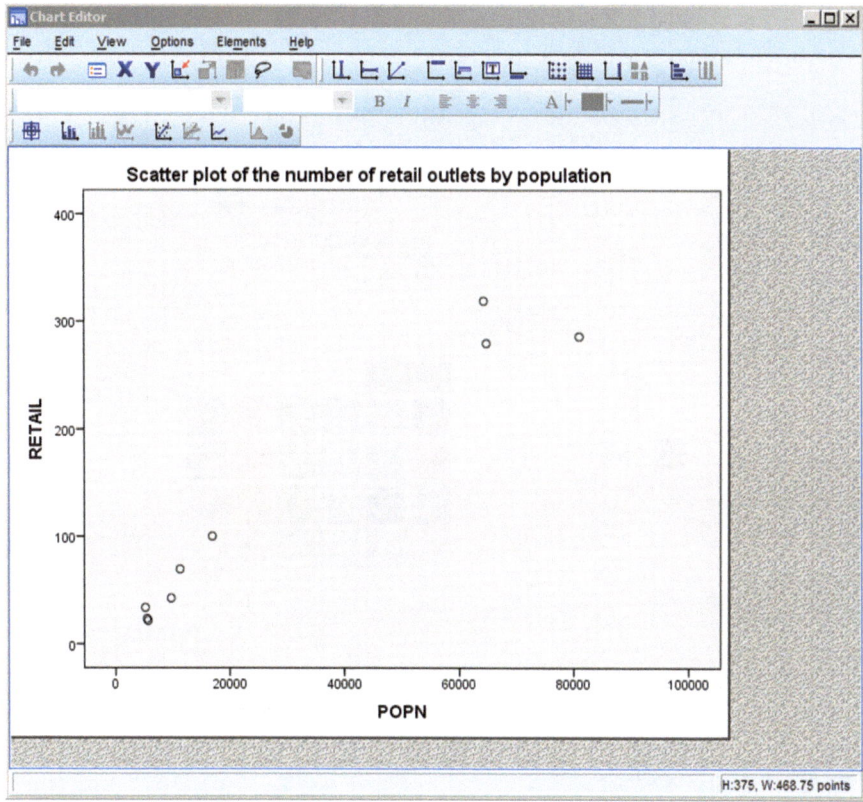

Fig. 1.15 The Chart Editor

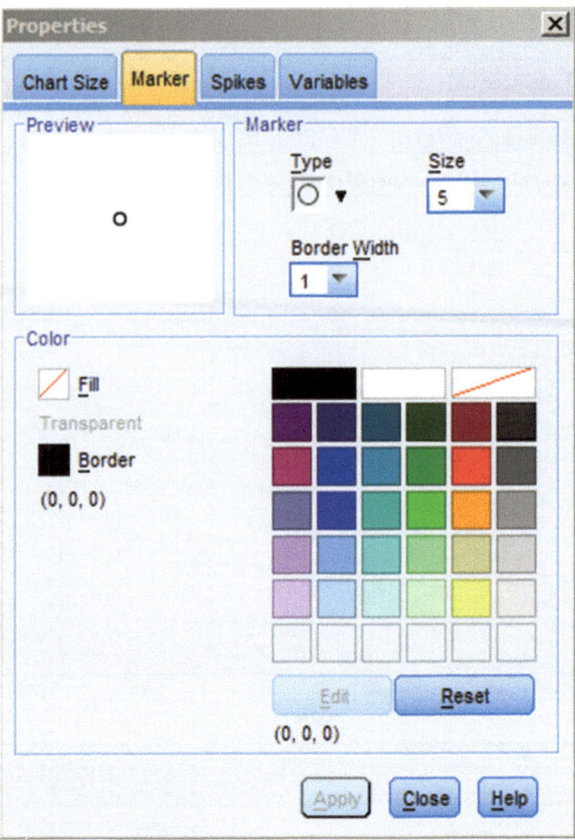

Fig. 1.16 The properties dialogue box

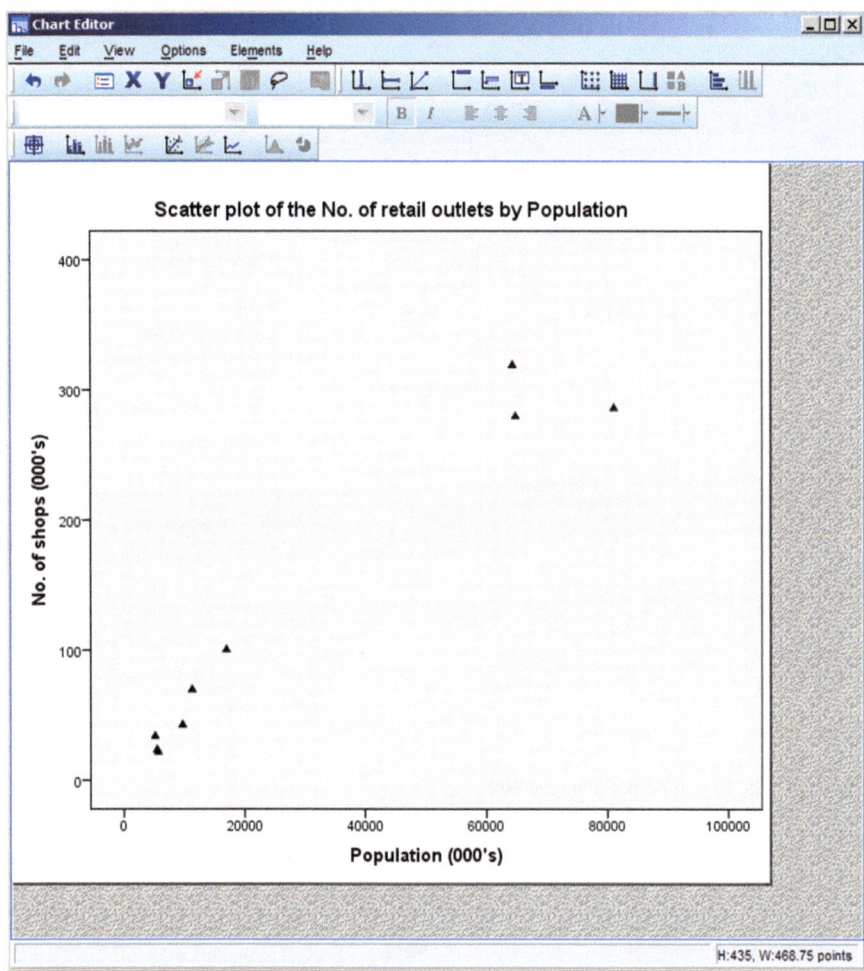

Fig. 1.17 The edited diagram in the Chart Editor

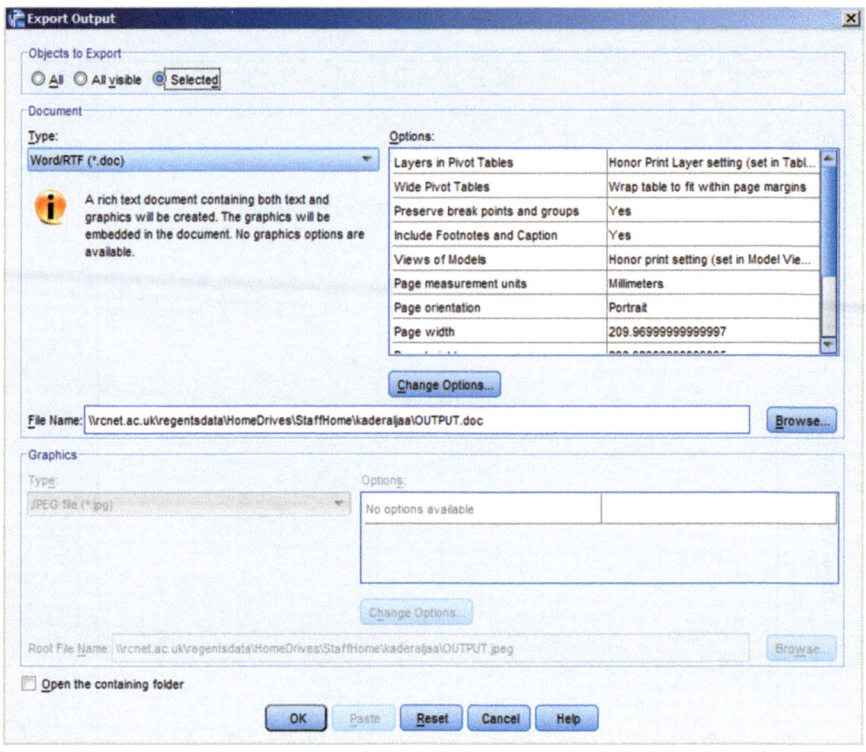

Fig. 1.18 The Export output dialogue box

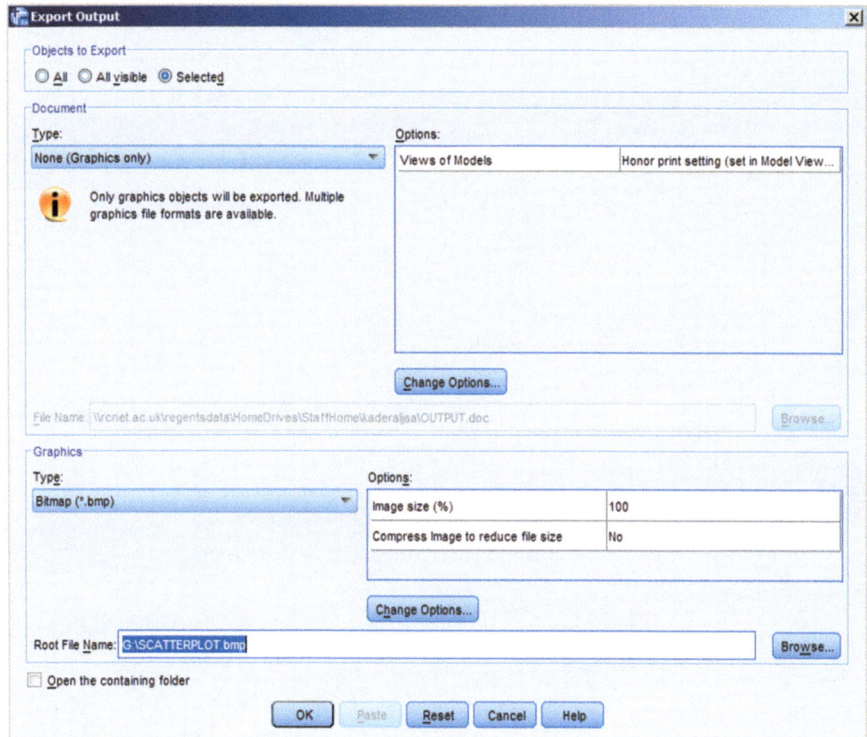

Fig. 1.19 The Export Output dialogue box: graphics output

Part II
Data Examination and Description

Chapter 2
Graphics and Introductory Statistical Analysis of Data

Much can be learned by constructing graphs and elementary statistical examination of the researcher's gathered data. Data input errors may be spotted and cases with extreme values highlighted. It is always relevant to examine the data at a basic level and in a systematic fashion. It is most certainly incorrect to search haphazardly for statistical significance. Based upon such rudimentary investigations, initial hypotheses may be modified or anticipated methods for further testing may have to be revised. The methods described in this chapter would often be precursors to the application of the techniques described in later chapters. The methods presented involve graphics and methods for initial data examination and description.

2.1 The Boxplot

Boxplots summarize the distributional characteristics of data but do not plot the raw data values themselves. Rather, they plot summary statistics for the distribution of the gathered data. The median is one statistic plotted on a boxplot. In Chap. 1, it was mentioned that the inter-quartile range is a measure of dispersion used in conjunction with the median and this range is also plotted on a boxplot.

The datum value below which 25 % of the gathered observations lie when the data are arranged in size order from lowest to highest, is called *the lower quartile*; that datum value below which 75 % of the observations lie when the data are arranged in size order is called *the upper quartile*. The difference between these two quartiles is *the inter-quartile range*. For example, if 25 % of the weights of an industrial product lie below 35.6 g. (the lower quartile) and 75 % of the weights lie below 43.9 g. (the upper quartile), then the inter-quartile range is 43.9–35.6=8.3 g. The middle 50 % of the data lies between the upper and lower quartiles, spanning 8.3 g.

Suppose we are studying factors that may have an effect on the growth of year-on-year revenue for a series of multinational firms. (We shall return to this idea in

© Springer International Publishing Switzerland 2016
A. Aljandali, *Quantitative Analysis and IBM® SPSS® Statistics*,
Statistics and Econometrics for Finance, DOI 10.1007/978-3-319-45528-0_2

the regression section of the present text). For the time being, however, we focus on basic data analysis involving just the variable named REVENUE which represents each company's yearly growth in revenue (based on 2015 revenue calculations). If you want to run through this exercise, the data file is named COMPANY REVENUE. SAV on the dedicated website page. The file contains data for n = 20 firms. Initially, we shall generate a boxplot of REVENUE. The boxplot is generated via:

Graphs
  Legacy Dialogs
    Boxplot…

which generates the Boxplot dialogue box of Fig. 2.1. Click the Define button to select a Simple Boxplot and 'Summaries of separate variables'. Enter REVENUE into the Variables box of the resultant dialogue box. Clicking OK generates the boxplot of Fig. 2.2.

The vertical axis represents data values of the variable that is subject of the boxplot i.e. REVENUE. The lower boundary of the shaded box is the lower quartile; the upper boundary is the upper quartile. The numerical values of these two quartiles may be read by reference to the vertical axis. Therefore, the LQ is around 0 % and the UQ about 25 %. Fifty percent of the observed cases have values within the box and the length of the box corresponds to the inter-quartile range. The black line inside the box represents the median value, which is here in the region of 5 %.

The IBM SPSS Statistics boxplot includes two categories of observations with extreme values. Cases (firms) with numerical values more than three box lengths from the upper or lower edge of the box are called *extreme values* and they are designated

Fig. 2.1 The Boxplot dialogue box

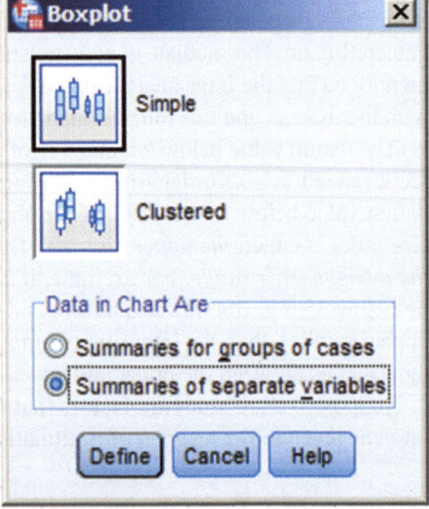

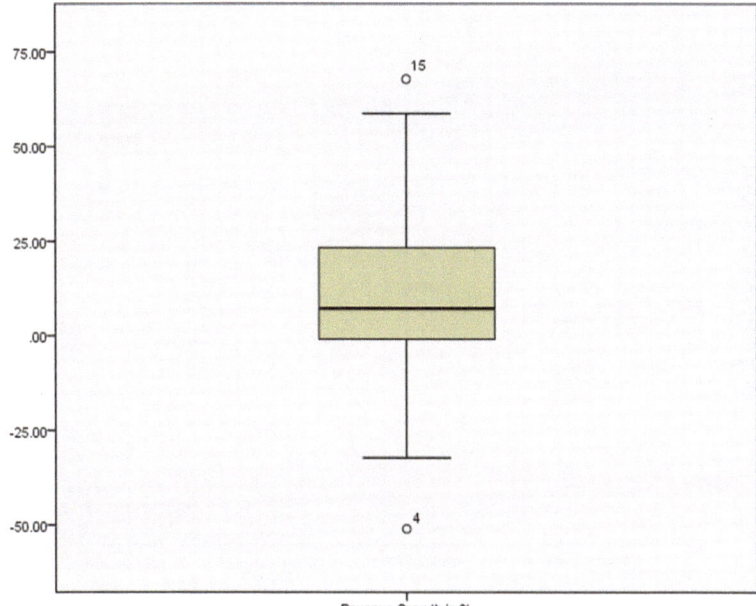

Fig. 2.2 A boxplot of a set of companies' revenue growth

with an asterisk (*) with the case number printed to the side. Cases with numerical values that are between 1.5 and 3 box lengths from the upper or lower edge of the box are called *outliers* and these are denoted by a circle, with the case number printed to the side. In Fig. 2.2, the 4th and 15th firm are outliers.

Lines are drawn from the upper and lower edges of the box to these values. These lines are called whiskers, which explains an alternative name for boxplots — *box-whisker plots*. The length of the box is indicative of the spread or variability inherent in the gathered data. If the median is not in the centre of the box, then the data must be skewed. For example, if the median is closer to the top of the box, then the data are negatively skewed. A principal use of boxplots is to compare the distributions of values in different groups. For example, we may wish to compare the distributions of company returns for firms of negative, low, middle and high return growth.

We can now group these 20 firms according to their revenue growth. The variable GROWTHGP (growth group) in the file COMPANY REVENUE.SAV labels the companies as '0' (NEGATIVE) if their revenue growth is negative, '1' (LOW) if their revenue is low, '2' (MEDIUM) if their growth was relatively good and a code of '3' (HIGH) if their growth is seen to be high. We can construct a boxplot where companies are classified based on their revenue growth labels. In the dialogue box of Fig. 2.1, select the option 'summaries for groups of cases' and click the Define button. Enter REVENUE into the 'Variables' box and GROWTHGP into the 'Category Axis' box. Clicking the OK button generates Fig. 2.3.

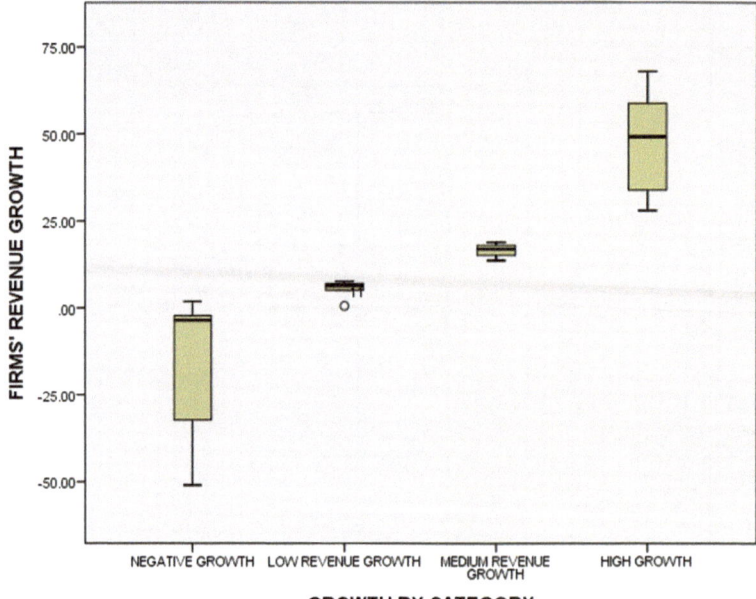

Fig. 2.3 Firms' REVENUE group by category

2.2   The Histogram

Several of the routines discussed in later chapters require that the gathered data are normally distributed, or at least do not deviate far from normality. (The normal distribution is sometimes referred to as the Gaussian distribution). A subjective method for assessing normality is the histogram. To generate a histogram for the variable REVENUE, select:

Graphs
  Legacy dialogs
    Histogram…

and the Histogram dialogue box of Fig. 2.4 is generated. Enter REVENUE into the Variable box. Also, it is wise to select the option 'display normal curve'. This superimposes the normal curve (inverted bell-shape) on the user's histogram and assists interpretation. Clicking the OK button generates Fig. 2.5.

Figure 2.5 is the result of editing the default IBM SPSS Statistics histogram in the Chart Editor. From the previous chapter, you may recall that you access the Chart Editor by double clicking inside the chart as it is displayed in the IBM SPSS Statistics Viewer. The default set up is that the histogram bars are an olive colour. To change this and to generate a plot similar to Fig. 2.5, click any bar of the histogram when in the Chart Editor. A blue surround indicates that all histogram bars are selected. Click the Show Properties button (3rd from the left). Select the Fill & Borders tab and via the Fill box, change the fill colour to white. Now select the Pattern box to select any

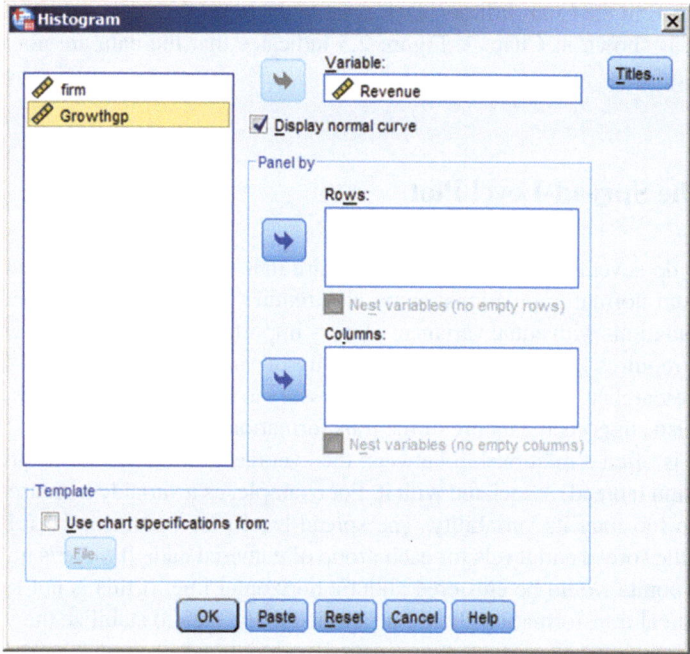

Fig. 2.4 The Histogram dialogue box

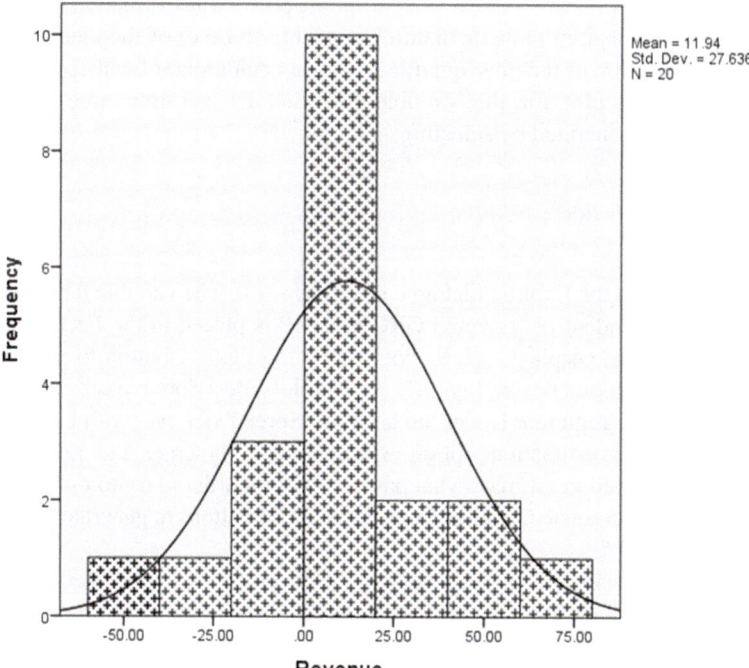

Fig. 2.5 A Histogram of firms' revenue growth

fill pattern. Exit the Chart Editor (black cross in the top right hand corner) and Export the chart as shown in Chap. 1. Figure 2.5 indicates that the data are not normally distributed.

2.3   The Spread-Level Plot

Not only do several statistical routines require that all groups of gathered data are drawn from normal populations; some also require that the data have been drawn from populations with equal variance. (A very important technique called *analysis of variance* requires both normality and equality of variances). The spread-level plot tells the researcher if a transformation is necessary to stabilize the variances and if so, the plot also suggests the nature of the transformation.

There is often a relationship between the average value (level) of a variable and the variation (spread) associated with it. For example, as a variable increases in magnitude, so too may its variability. The spread-level plot examines the relationship between the spread and levels for each group of gathered data. If there is no relationship, the points would be clustered about a horizontal line. If this is not the case, a mathematical transformation should be applied to the data to stabilize the variances, if this is a required property of the statistical technique to be used. The spread-level plot suggests power transformation. A power of 3 would cube all the data values; a power of 0.5 would root all the data values. A power transformation of 1 results in no change in the data values. To assess an appropriate power transformation for the data, the spread-level diagram plots the natural logarithm (base e) of the median against the natural logarithm of the inter-quartile range for each group of gathered data.

A spread-level plot for the variable REVENUE over the three classes of GROWTHGP is generated by selecting:

Analyze
   Descriptive Statistics
      Explore…

which gives rise to the Explore dialogue box of Fig. 2.6. The variable REVENUE is entered into Dependent list box and GROWTHGP is placed in the Factor list box. Under the heading 'Display', select Plots. Click the Plots… button to produce the Explore: Plots dialogue box of Fig. 2.7. No boxplots, descriptive plots or normality plots with tests are required. Under the heading 'Spread vs Level with Levene Test' choose the Power estimation option, since we simply enter the raw data for REVENUE and need to estimate what power to take in order to try to make the variances of each group equal. Click the Continue and OK buttons to generate the spread-level plot of Fig. 2.8.

The filled in triangular markers in Fig. 2.8 were obtained in the IBM SPSS Statistics Chart Editor—the default being unfilled circles. There are four groups in Fig. 2.8, it appears that the points of three groups do congregate around a horizontal line, so any assumption about the equality of variances in the three groups seems not

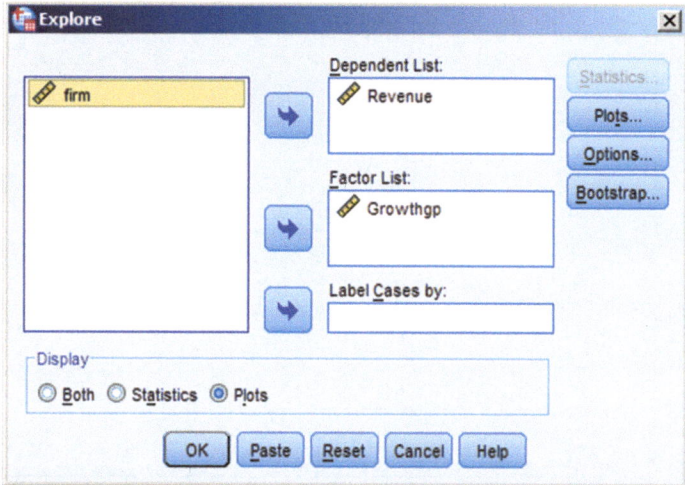

Fig. 2.6 The Explore dialogue box

Fig. 2.7 The Explore:
Plots dialogue box

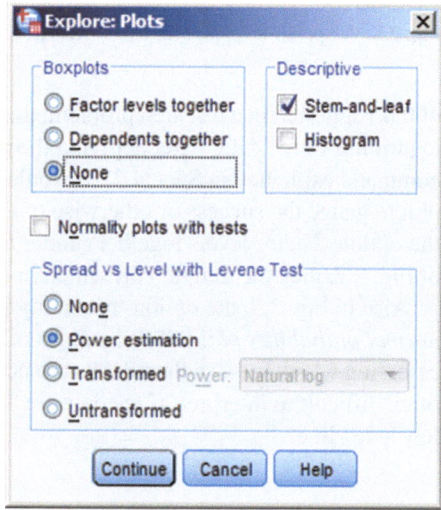

to be dubious. From the slope of a line drawn through the middle of these four points, IBM SPSS Statistics estimates the power value that will transform the data and bring us closer to the situation of equal variances. The appropriate power is obtained by subtracting the slope of this line from unity. Here, a power transformation of $1-(1.124)=-0.124$ is suggested. This power is virtually unity, which suggests that the data should not be transformed. This apparent contradiction is due to having only a small number of groups upon which to base a decision. If a power transformation

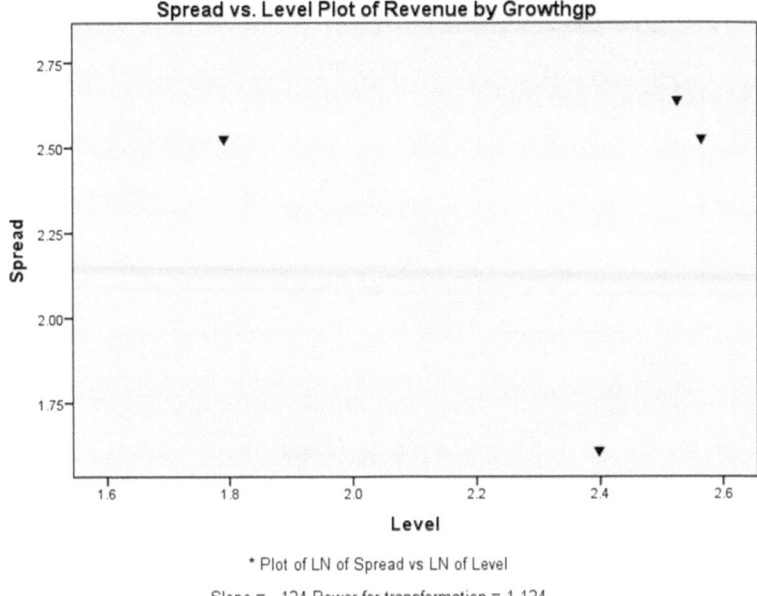

Fig. 2.8 A Spread-level plot for firms' REVENUE

of 0 is suggested in a research problem, this is in fact interpreted as suggesting taking logarithms of the data. After applying the power transformation (via the COMPUTE command explained in Sect. 4.2), it would be prudent to obtain another spread-level plot to gauge the success or otherwise of the transformation. If the user had selected the option 'Factor levels together' under the 'Boxplot' heading in the dialogue box of Fig. 2.7, then the diagram presented in Fig. 2.3 would have been the result.

Also in Fig. 2.7, the option 'normality plots with tests' generates what is called a *normal probability plot* (NPP) amongst others. NPP is another graphical method for assessing whether the gathered data are normally distributed, but these diagrams are often difficult to interpret. If the data are normal, then the trace in a NPP should follow a 45° line.

2.4   Bar Charts

The file LABOUR.SAV contains an estimate employment data (in thousands) for the 11 regions of England, Scotland and Wales at June 2008. Besides the string variable REGION, there are three variables in this file and their IBM SPSS Statistics names are in capitals below

- TOTAL—total no. of employees in all industries
- WHOLEDIS—no. of employees in wholesale distribution
- RETDIST—no. of employees in retail distribution.

These variables have already been labelled and defined in LABOUR.SAV. In IBM SPSS Statistics, it is possible to construct simple, clustered and stacked bar charts. These charts can act as summaries for groups of cases or separate variables, as well as represent the values of individual cases. Here, we shall construct a clustered bar chart of WHOLEDIS and RETDIST as defined above. The chart will represent the number of employees in retail and wholesale distribution for each of the 11 geographical regions. The type of bar chart generated depends on the manner in which the data file has been organized. Available options are:

- Summaries for groups of cases—cases are counted, or one variable is summarized in subgroups. For example, one could plot SALES over 11 REGIONS, where the regions are coded from 1 to 11.
- Summaries of separate variables—more than one variable is summarized. Each variable could be summarized within categories of another variable. For example, we may plot the sales of three products PROD1, PROD2 and PROD3 over the 6 years between 2007 and 2012.
- Values of individual cases—individual values of one or more variable are plotted over the cases in the data file. This is what is required here, namely a plot of WHOLEDIS and RETDIST for each of the cases, which are the 11 geographical regions.

Assuming that you have opened LABOUR.SAV, from the Data Editor, click:

Graphs
  Legacy dialogs
    Bar...

to produce the Bar Charts dialogue box of Fig. 2.9. Click the Clustered option (an outline box will appear around this selection) and the data in the chart are 'values of individual cases' (at the bottom of Fig. 2.9). Now click the Define button. We now have the Define Clustered Bar dialogue box of Fig. 2.10 and in which the minimum specifications are a category axis variable and two or more bar variables. From the list of variables on the left hand side, select (click) WHOLEDIS and RETDIST and click the top arrow to insert them into the box headed by 'Bars Represent'. Now select (click) the variable REGION and in the box titled 'Category Labels', click 'Variable' and the arrow button to enter this variable as the category (horizontal) axis variable.

If we had not selected 'Variable', then by default, the Case Number would have appeared on the horizontal axis. Click OK to generate the clustered bar chart, which is shown in Fig. 2.11.

Figure 2.11 was edited by entering the IBM SPSS Statistics Chart Editor. To change the shading in the bars, select the bars one at a time from the legend. If you point to the bars inside the chart, all bars—green and blue—will be selected. Use the Show Properties Window button together with the Fill and Pattern options to change the shading style. The default label for the vertical axis is 'Value'. To change this, point at the word 'Value', click twice and it will appear horizontally for editing. After the first click, you will see the blue surround, indicating that it has been selected. Erase the word 'Value' and replace it as shown. Hit the Enter key when completed. You can point to the X and/or Y axis titles and change the fonts and point sizes.

Fig. 2.9 The Bar Charts
dialogue box

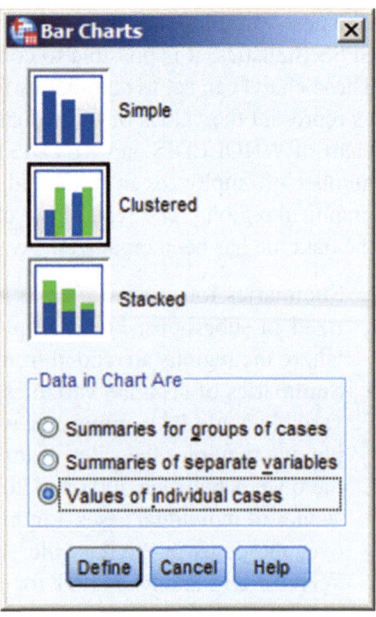

Figure 2.12 is a stacked version of Fig. 2.11. It is obtained by making the relevant selection in the dialogue box presented in Fig. 2.9.

2.5   Pie Charts

Pie charts may be constructed to summarize groups of cases or separate variables and the values of individual cases. Using the data file LABOUR.SAV, we are going to use this latter option to draw a pie chart of the variable TOTAL (IBM SPSS Statistics variable name 'Total Number of Employees in all Industries') by REGION.

To obtain a pie chart, from the menu choose:

Graphs
   Legacy dialogs
      Pie…

which generates the Pie Charts dialogue box of Fig. 2.13. Click the option 'Values of individual cases' and click the Define button to produce the Define Pie dialogue box of Fig. 2.14. The 'Slices' are to represent the variable TOTAL, which is clicked and entered into the appropriate box. We will label the slices by the names of each REGION. Under the heading 'Slice Labels', click 'Variable' and enter REGION via the arrow button. Click OK to generate the pie chart. After setting a title, the resultant chart is shown in Fig. 2.15.

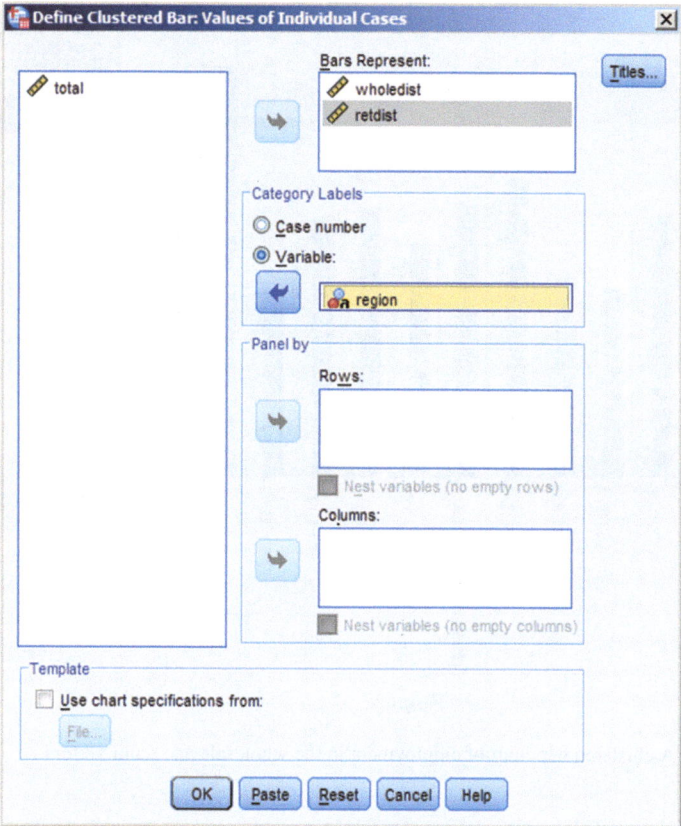

Fig. 2.10   The Define Clustered Bar Charts dialogue box

Figure 2.15 was obtained by editing the default chart. As usual, double click to enter the IBM SPSS Statistics Chart Editor. Click the Show properties Window button to change the fill patterns for the pie chart slices via the Fill and Pattern options. It is useful to add information to the slices. The ▥ icon permits the percentages of observations in each slice to be displayed. Clicking this icon gives rise to the Properties dialogue box of Fig. 2.16. Under the 'Displayed', one can see that the percentages of employees in each region are to be displayed. If the user wants the raw numbers, then click Total Employees in all Regions in the 'Not Displayed' box and use the upward pointing arrow to insert this variable into the 'Displayed' box. You can change the location of any displayed labels under the heading 'Label Position'.

If one or more of the regions is a special focus in a study, then it is possible to "explode" the relevant slice for emphasis. Click the slice to be exploded—I have chosen Scotland. You will see the blue surround, indicating that the slice has been selected. ◖ Click the icon while in the Chart Editor. Figure 2.17 presents the resultant pie chart.

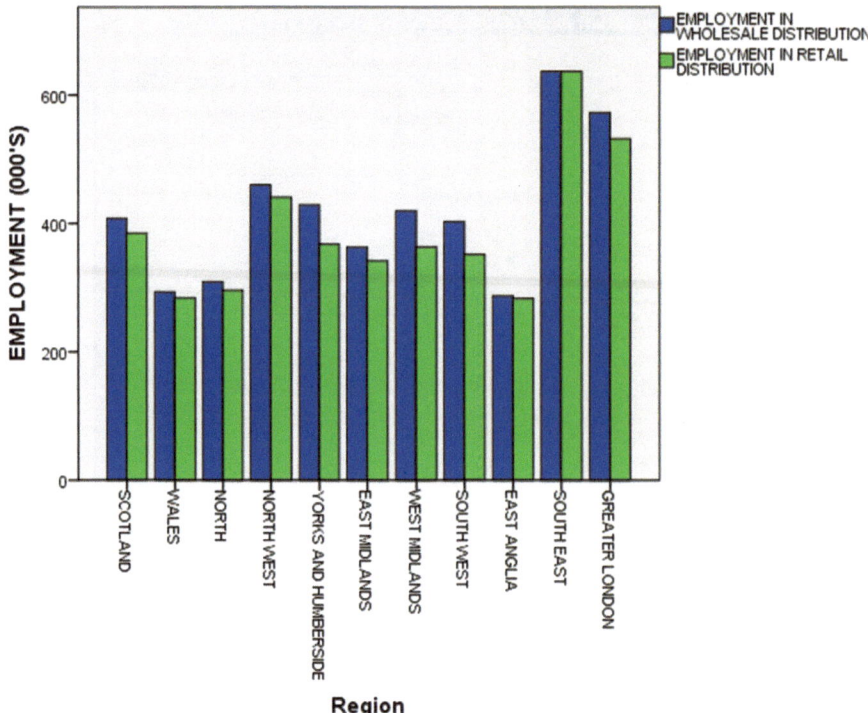

Fig. 2.11 A clustered bar chart of employment in the wholesale and retail sectors

2.6   Pareto Charts

A Pareto chart is a bar chart in which the cases are sorted in descending order. A curve is usually added to show the cumulative frequency and percentages across cases. Pareto charts are used widely in quality control exercises, in that they focus on the most important category out of all possibilities. We shall construct a Pareto chart for the number of employees in the retail trade (RETDIST) by REGION. From the IBM SPSS Statistics Data Editor, click:

Analyze
    Quality Control
        Pareto Charts…

which produces the Pareto Charts dialogue box of Fig. 2.18. Here, we shall construct a simple Pareto chart (click) and the regional data are 'Values of individual cases' (click). Upon clicking the Define button we produce the 'Define Simple Pareto' dialogue box of Fig. 2.19. Click the variable RETDIST into the 'Values box' via the top arrow button. We shall use the variable REGION as the variable to label the cases as shown in the resultant plot presented in Fig. 2.20. As usual, the fill pattern for the bars has been modified in the IBM SPSS Statistics Chart Editor.

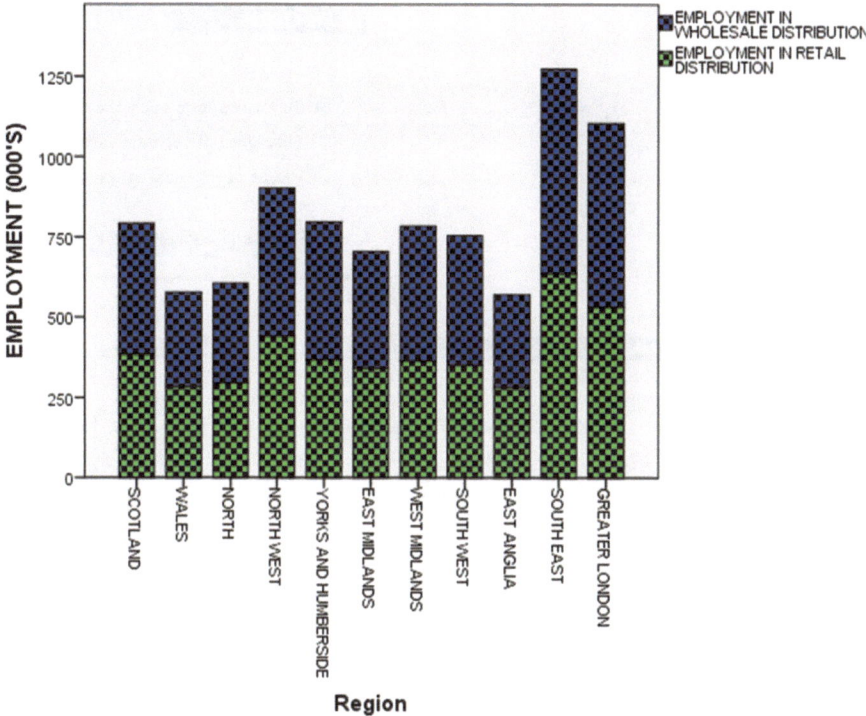

Fig. 2.12   A stacked bar chart of employment in the wholesale and retail sectors

Note that the regions are sorted from the highest level of retail employment to the lowest. The right vertical axis represents cumulative percentage. The left vertical axis represents cumulative frequency. The actual numbers of employees are entered by default in each bar. A stacked Pareto chart could be produced to show the cumulative frequencies and percentages of employees in the wholesale and retail sectors combined. This is achieved by choosing the option 'Stacked' in Fig. 2.18. In the resultant Define Stacked Pareto dialogue box, the variables RETDIST and WEHOLEDIS are entered into the 'Values' box.

The stacked Pareto chart is shown in Fig. 2.21.

2.7   The Drop-Line Chart

The essence of a drop-line chart is that it shows the difference between two variables—for example, the numbers of employees in wholesale distribution and retail distribution. In that the former figures are greater than the latter, the wholesale distribution figures will appear above those representing the retail distribution. Click:

Fig. 2.13 The Pie Charts
dialogue box

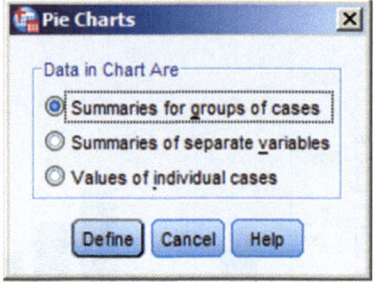

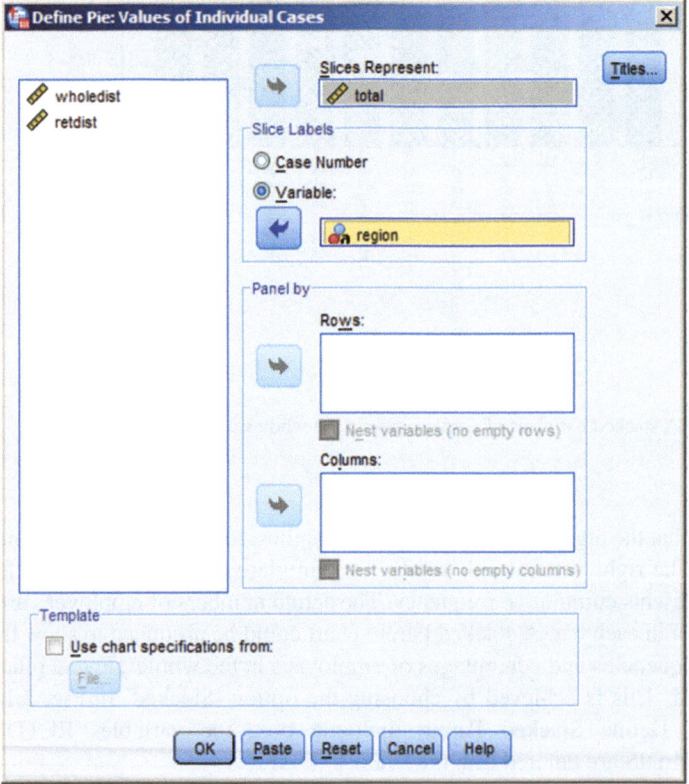

Fig. 2.14 The Define Pie dialogue box

Graphs
  Legacy dialogs
    Line…

which generates the Line Charts dialogue box of Fig. 2.22. Select the 'Drop-Line'
option and choose 'values of individual cases' (i.e. the regions) at the bottom of
Fig. 2.22. The dialogue box of Fig. 2.23 is now produced. Place the two variables in

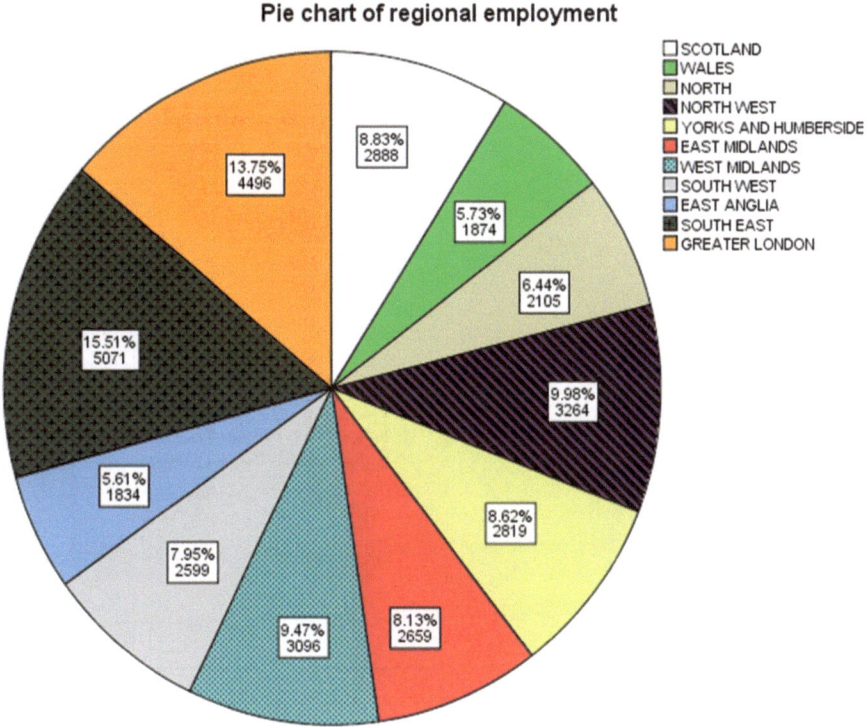

Fig. 2.15 The resultant pie chart

the 'Points Represent' box and select the region as the 'Variable' under 'Category Label'. Click the OK button to produce Fig. 2.24.

Note that the number of employees in wholesale and retail are equal for the south east region. Naturally, it is possible to edit the symbols used on this plot in the IBM SPSS Statistics Chart Editor.

2.8 Line Charts

To construct a line chart, click:

Graphs
    Legacy dialogs
        Line…

from the Data Editor. This opens the Line Charts dialogue box of Fig. 2.22. We are going to construct a line chart that shows both the numbers of employees in the

Fig. 2.16 The pie chart
Properties dialogue box

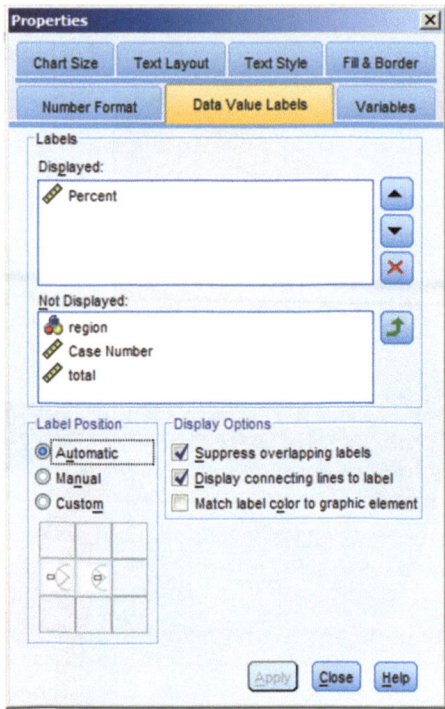

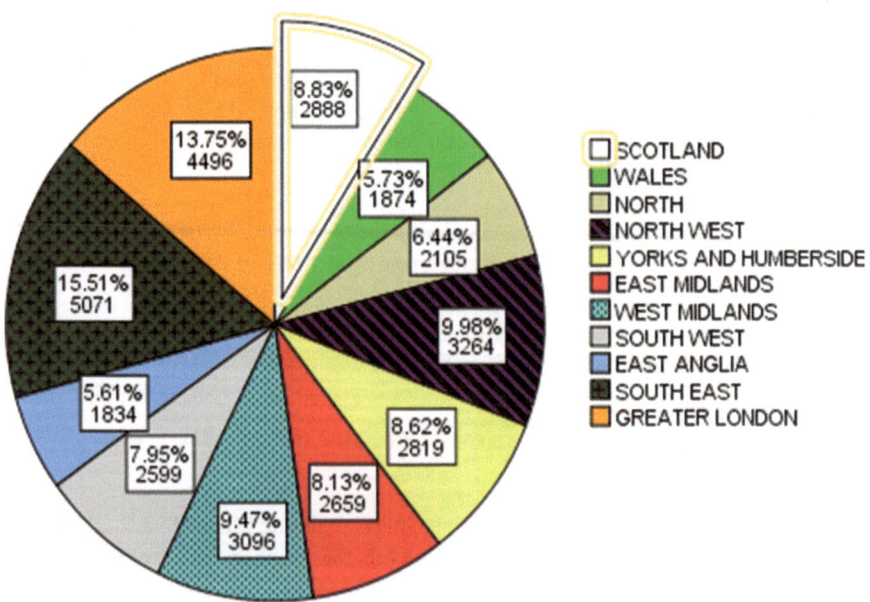

Pie chart of regional employment

Fig. 2.17 A pie chart with an exploded slice

Fig. 2.18 The Pareto
Charts dialogue box

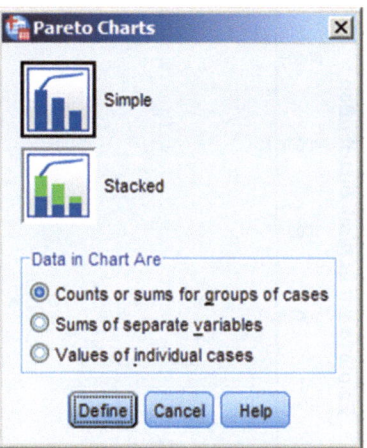

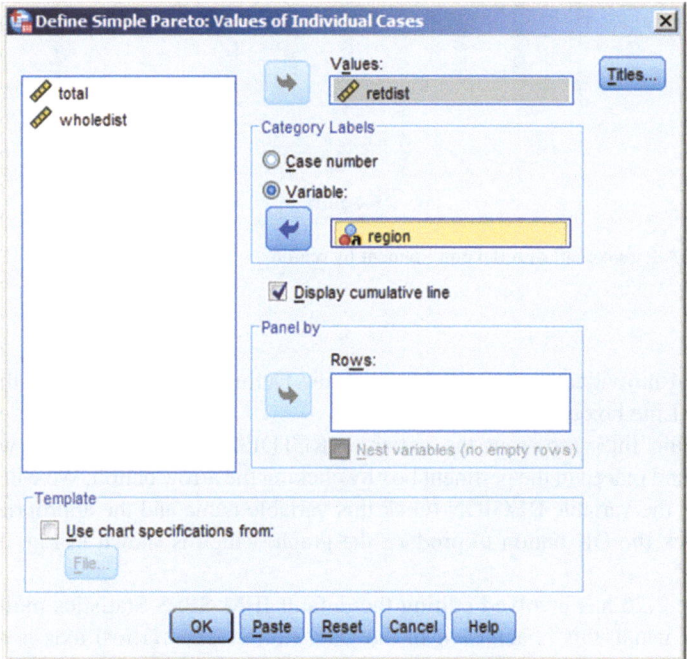

Fig. 2.19 Define Simple Pareto dialogue box

retail (REDIST) and wholesale (WHOLEDIS) trades. Drawing more than one variable gives rise to multiple line chart, drawing a single variable involves a simple
line chart. In the Line Charts dialogue box click the option 'Multiple' and a dark
surround will appear to indicate this selection. We are going to record the number of
employees in both trades by geographical region. Therefore, the data in the chart are

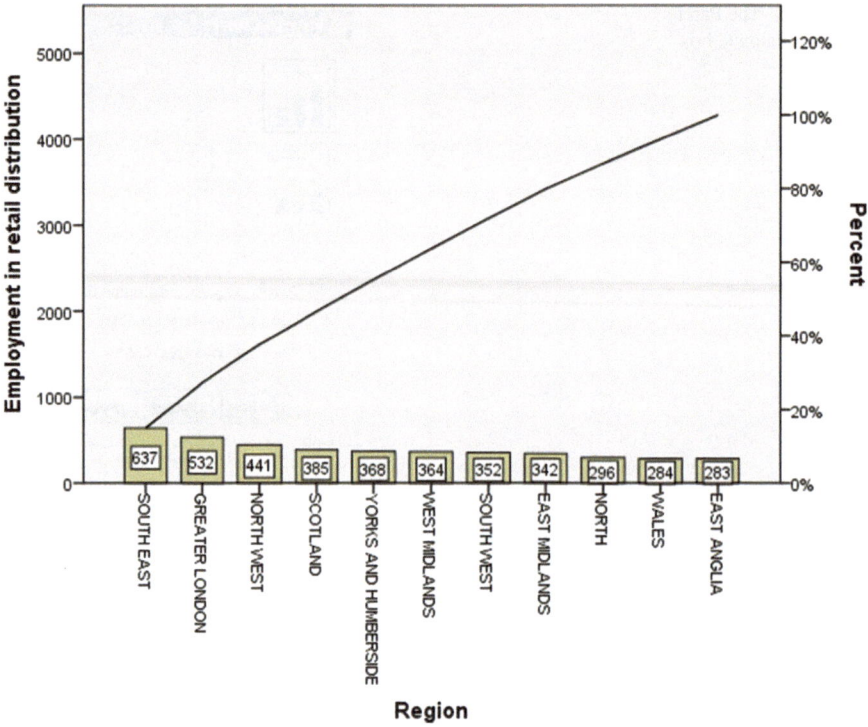

Fig. 2.20 A Pareto chart of retail employment by region

'Values of individual cases' (click). Click the Define button to generate the Define Multiple Line box of Fig. 2.25.

Here the lines represent the variables RETDIST and WHOLEDIS which are selected and placed in the pertinent box by clicking the arrow button. We will label the data with the variable REGION (click this variable name and the appropriate arrow key). Click the OK button to produce the graph, which is shown in Fig. 2.26 after editing.

Figure 2.26 has involved editing the default IBM SPSS Statistics multiple line chart. As usual, this is achieved in the Data Editor. The vertical axis is given the default title of 'Value'. This has been changed in the manner described previously. The default lines are coloured green and blue. In the legend, click one of the two lines, then the Show Properties Window button. This gives rise to the dialogue box of Fig. 2.27 and I have changed the colour to black. Under the heading 'Lines', I have changed the Style of the line representing employment in the wholesale sector into that of a broken line. I have left a solid line for employment in the retail sector line. The Weight in Fig. 2.27 refers to the thickness of the lines. I have changed the default of 1.0–1.5 for both lines, thereby increasing the thickness.

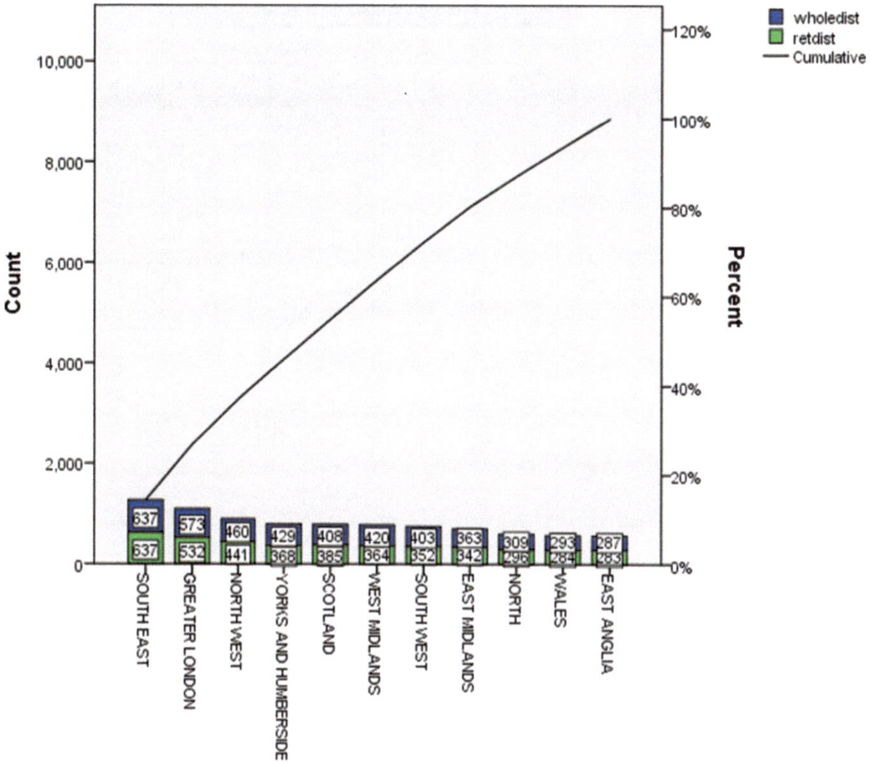

Fig. 2.21 A stacked Pareto chart of employment in the retail and wholesale sectors

Fig. 2.22 The Line Charts
dialogue box

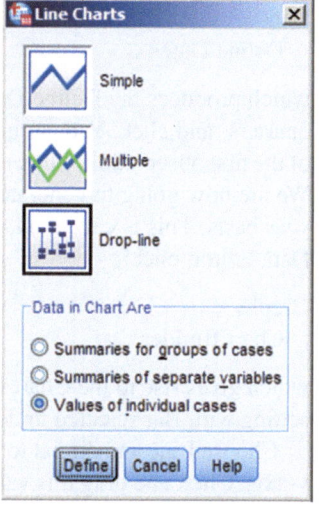

Fig. 2.23 The Define
Drop-line dialogue box

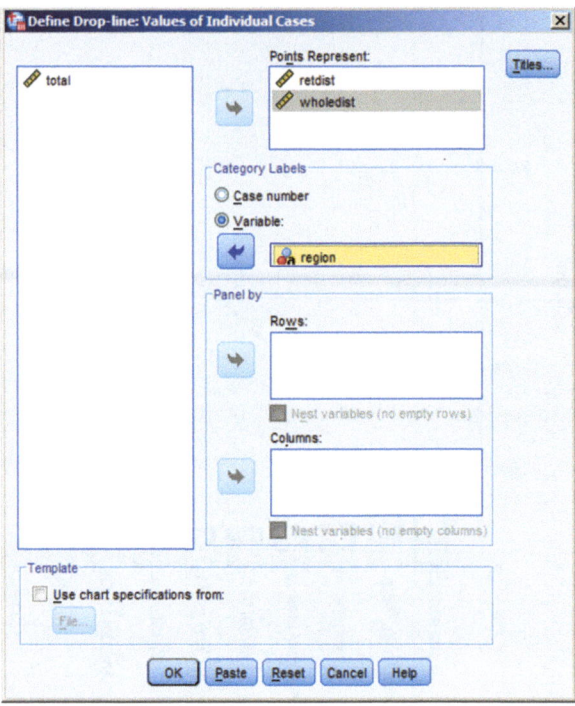

2.9   Applying Panelling to Graphs

AMERICAN TOURISM.SAV contains UK earnings predictions (£ millions at current prices) from American tourists on a quarterly basis, starting at 2007 Q1. The variable name is NAMERICA. The data are not dated. To ascribe dates to the observations, click:

Data
   Define Dates…

which produces the Define Dates dialogue box. Scroll up to find the option 'Years, quarters' and click. You are now asked to define the 'Year' (2007) and 'Quarter' (1) of the first observation. Click the OK button and the dated file will appear as Fig. 2.28. We are now going to generate line charts showing quarterly earnings on a year-by-year basis. This is called "*panelling the data by year*". In the IBM SPSS Statistics Data Editor, click:

Graphs
   Chart Builder…

which gives rise to the Chart Builder dialogue box of Fig. 2.29 which is the default setting with Bar selected under the Gallery tab.

   Choose Line and drag it to the chart preview space. We want NAMERICA up the y-axis. Click and drag this variable from the variable list and insert it up the y-axis.

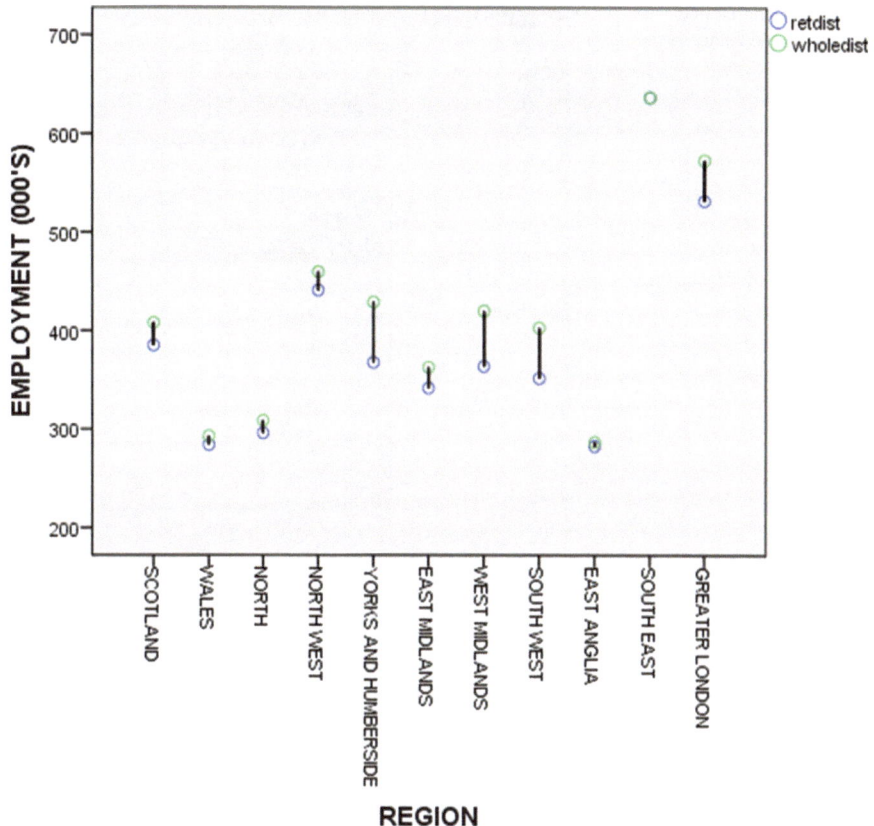

Fig. 2.24   A drop-line chart of regional employment in the retail and wholesale sectors

Click and drag the variable QUARTER_ and insert it in the x-axis. We are going to panel on an annual basis, so click the Groups/Point ID tab and tick the box Rows panel variable as shown in Fig. 3.1. Now drag the variable YEAR_ into the box titled 'Panel?'. Clicking the OK button generates Fig. 3.2 which has been edited in the Chart Editor.

The purpose of panelling is to examine changing behaviour over time. The user has to make the decision as to which is an appropriate panel variable within the context of the conducted research. Figure 3.3 shows the same information except that bar charts are used. It is obtained by the same process, except that initially the selection is:

Graphs
  Chart Builder…

and choose Bar under the Gallery tab. Follow the same steps as per Fig. 3.1.

Fig. 2.25 The Define
Multiple Line Chart
dialogue box

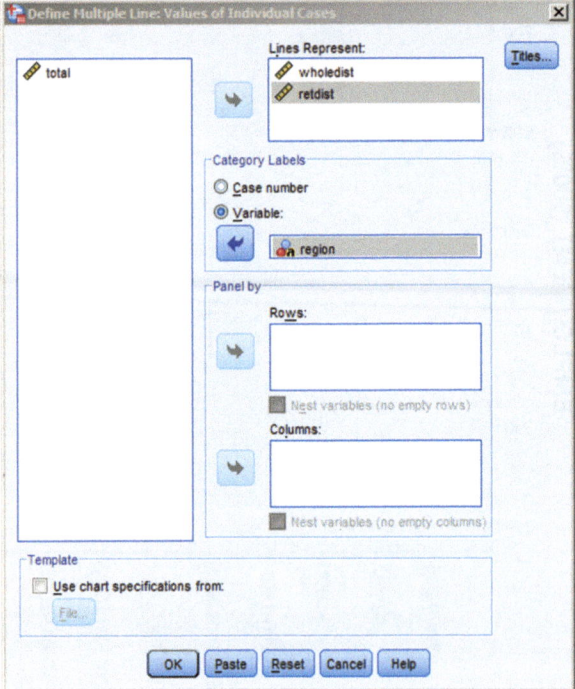

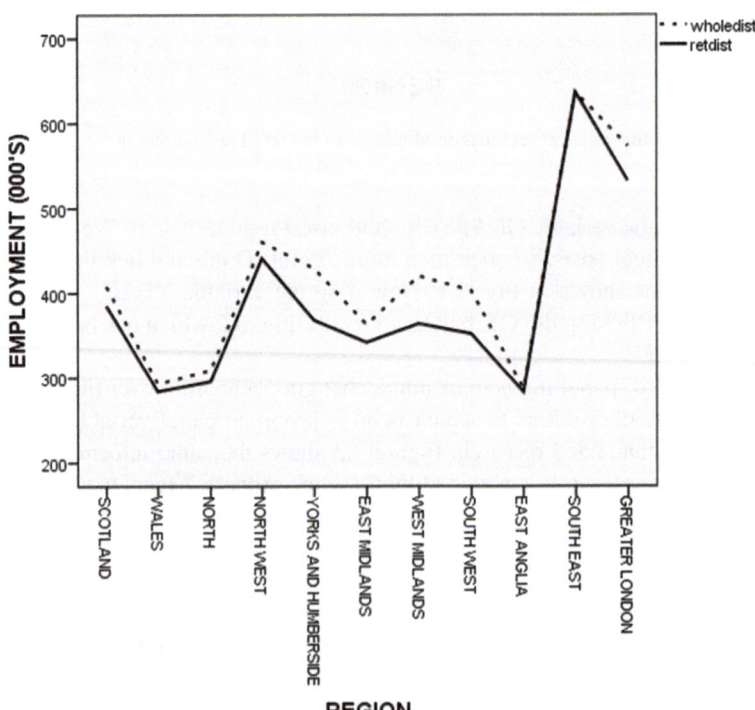

Fig. 2.26 A multiple line chart of employment in the retail and wholesale sectors

Fig. 2.27 The Properties
dialogue box for editing
lines

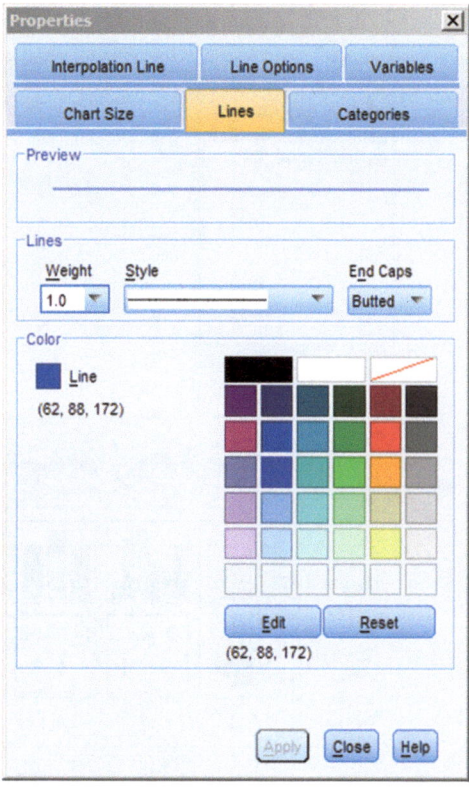

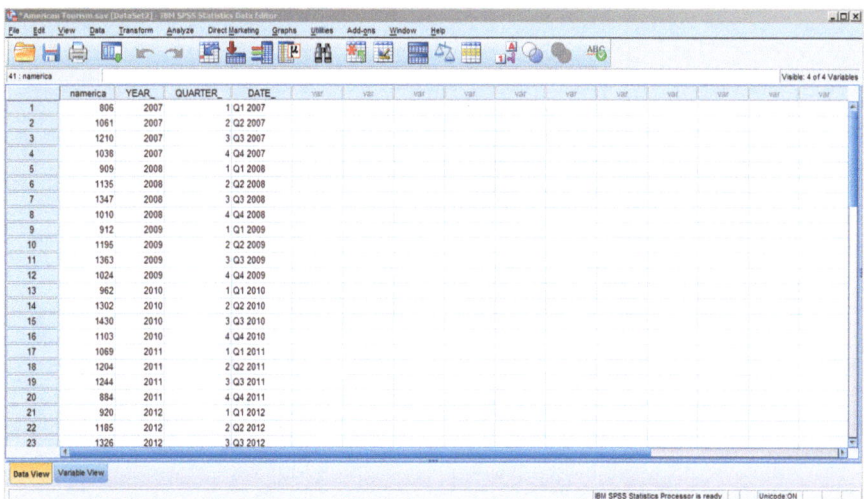

Fig. 2.28 UK tourism earning from American visitors

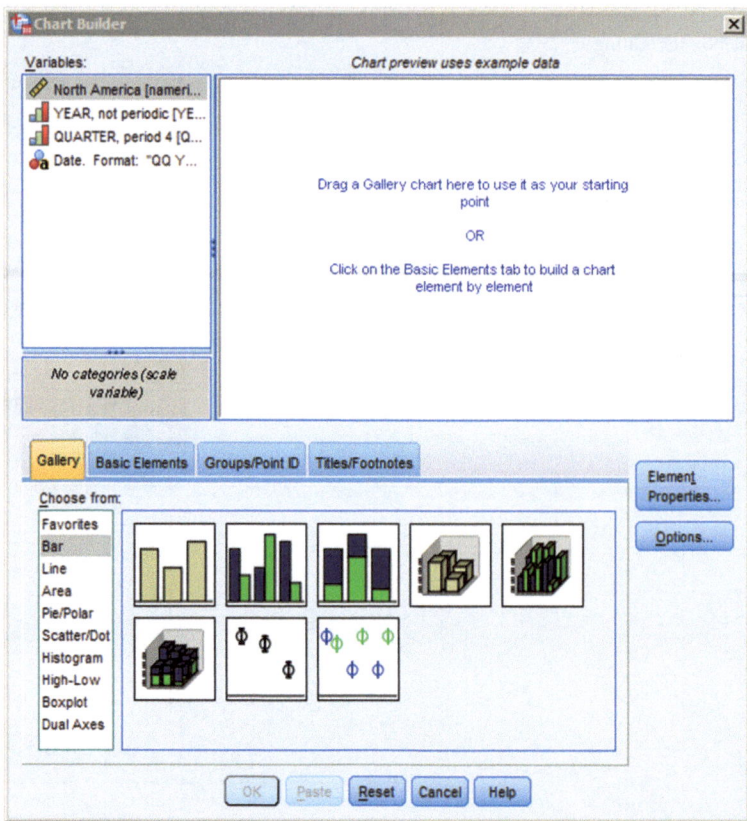

Fig. 2.29 The Chart Builder dialogue box

Chapter 3
Frequencies and Crosstabulations

3.1 Data Exploration via the EXPLORE Routine

Graphics readily communicate the characteristics of data if correctly and appropriately applied. However, there is sometimes subjectivity involved in interpretation. For example, it is not always clear whether a histogram suggests normality of the variable in question or whether a spread-level plot indicates that the variance of a variable is constant across various groups. In such instances of ambiguity, the researcher needs less subjective methods (Figs. 3.1, 3.2 and 3.3).

Reopen the file called COMPANY REVENUE.SAV. We are going to explore the variable REVENUE over the above three previously defined categories of GROWTHGP. To access the EXPLORE routine in IBM SPSS Statistics, click:

Analyze
  Descriptive Statistics
    Explore…

this produces the Explore dialogue box that we saw in Fig. 2.6. Click the variable REVENUE and click the top arrow to place this variable in the 'Dependent List'. Now click the variable GROWTHGP and click the arrow to place it in the 'Factor List'.

At the right hand side of the Explore dialogue box are Statistics and Plots that may be selected. Click the Statistics… button to obtain the Explore: Statistics dialogue box of Fig. 3.4. Choose the option 'Descriptive Statistics' (such as the mean, standard deviation, skewness etc.). Click the Continue button to return to the Explore dialogue box and then click the OK button. The results in Fig. 3.5 are generated.

© Springer International Publishing Switzerland 2016   53
A. Aljandali, *Quantitative Analysis and IBM® SPSS® Statistics*,
Statistics and Econometrics for Finance, DOI 10.1007/978-3-319-45528-0_3

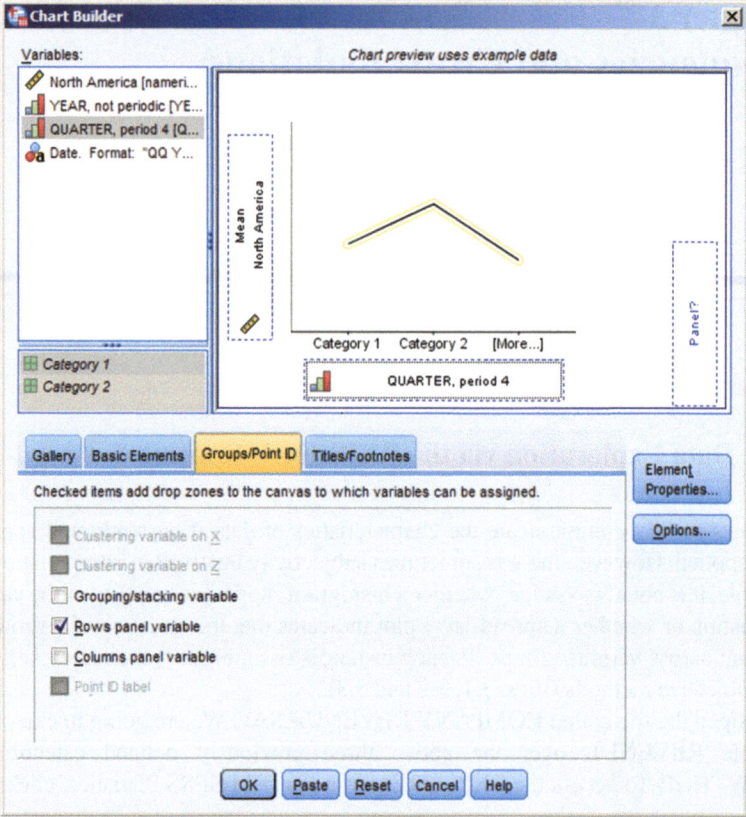

Fig. 3.1 Defining a panel variable in the Chart Builder

3.2   Statistical Output from EXPLORE

For brevity, descriptive statistics related to the variable REVENUE are reported only for only the group of firms with negative or low return growth in Fig. 3.5. There were 11 firms in these two groups. Under the statistics reported, the '5 % Trim' refers to the 5 % trimmed mean. The *trimmed mean* is an example of what is called a *robust estimator of location*. The 5 % trimmed mean eliminates the highest and lowest 5 % of observations. This estimate of location is, therefore, based on 90 % of the gathered data values. The advantage of the trimmed mean is that it is not influenced by any possible extreme values. It is based on a much larger set of middle values, unlike the median.

The trimmed mean makes better use of the data than does the median. (The latter statistic may be regarded as a 50 % trimmed mean). Further initial data investigation may be obtained by clicking the Plots… button in the Explore dialogue box of Fig. 2.6, which gives rise to the Explore: Plots dialogue box that was presented in Fig. 2.7. Unlike the last time we visited the latter dialogue box, select the option 'Normality

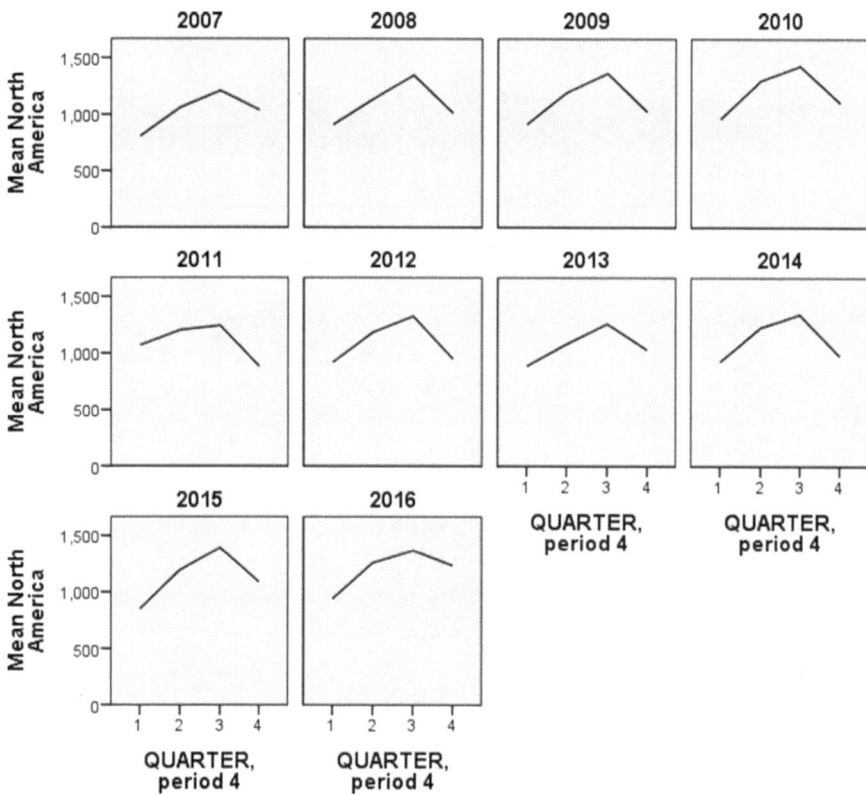

Fig. 3.2 UK quarterly earnings from American tourism panelled on an annual basis: line charts

Plots with Tests'. Also, under the heading 'Spread vs Level with Levene Test' select the option 'Power Estimation'; click the Continue button then the OK button.

As part of the output presented in the IBM SPSS Statistics Viewer, the *Shapiro-Wilks* and *Kolmogorov-Smirnov statistics* presented in Fig. 3.6 provide less subjective assessments of normality than do graphs such as histograms. Either may be used to assess if the data are normally distributed. Associated with these statistics in Fig. 3.6 are what is referred to as levels of *significance*. The concept of significance is discussed in more detail in Chap. 5. Suffice it to say at this point and in general, that the significance associated with a statistic is:

> The probability that a statistical result as extreme as the one observed would occur if the assumption under which the statistic was calculated is true.

For example, both of the Shapiro-Wilks (S-W) and Kolmogorov-Smirnov (K-S) statistics are computed under the assumption that the data are normally distributed. Examination of the above table for firms classified as having medium revenue growth indicates that there is a probability of 0.720 of obtaining a S-W statistic as extreme as 0.951 under the assumption that the returns are normally distributed. Low values

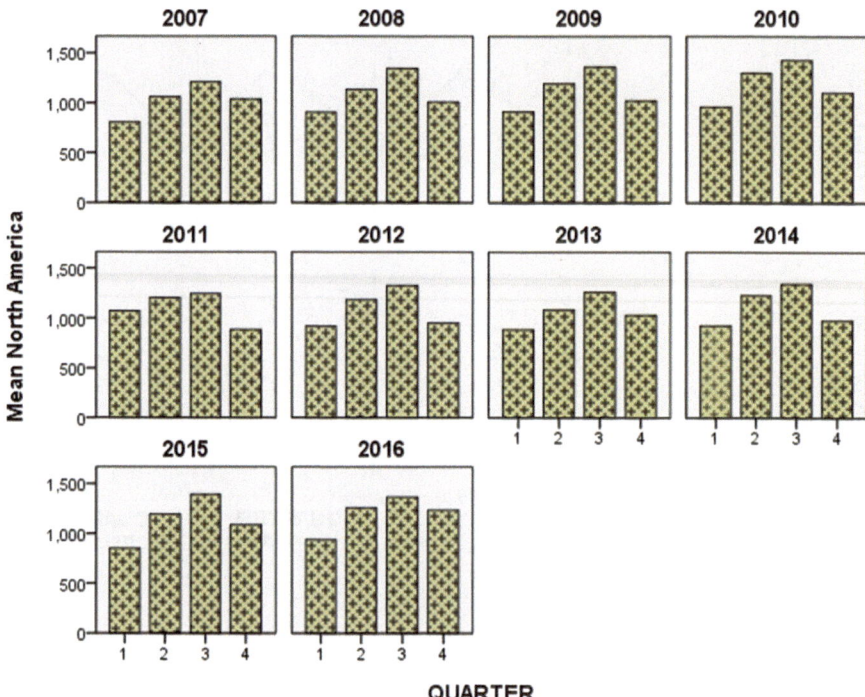

Fig. 3.3 UK quarterly earnings from American tourism panelled on an annual basis: bar charts

for the significance would cause us to doubt that the data are normal. The levels of significance of 0.720 (for S-W) and >0.200 (for K-W) are not so low as to refute normality. Conventionally, significance levels of less than 0.05 or sometimes 0.01 for these two statistics would be so low as to cast doubt on the idea of normality of the returns. Also as part of the IBM SPSS Statistics output, the *Levene statistic* is also reported as per Fig. 3.7.

The Levene statistic is computed under the assumption that REVENUE for the four groups of firms are drawn from populations with equal variances. The property of equal variances is often referred to as *homogeneity of variances*. There are alternative statistical methods available to assess the equality of variance, but competing methods tend to depend on the data being derived from normal populations. The Levene statistic is less dependent on the normality assumption. In Fig. 3.7 under the heading 'Based on Mean', the Levene statistic has numerical value 8.011 with significance level 0.02. The significance associated with this statistic indicates that there is a probability of 0.02 of obtaining a value as extreme as 8.011 for the Levene statistic under the assumption of equality of variance. Conventionally, a significance of below 0.05 would suggest rejection of the above notion. We, therefore, reject the idea that the REVENUE for the FOUR groups of firms are drawn from populations with equal variance. We have sufficient evidence that the spread of REVENUE for the four groups of firms is not the same. The results of this section illustrate that useful as graphical plots may be in assessing gross violations of assumptions like equality

Fig. 3.4 The Explore:
Statistics dialogue box

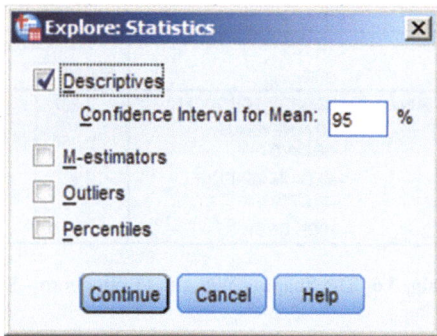

Descriptives

Growthgp				Statistic	Std. Error
Revenue	NEGATIVE GROWTH	Mean		-15.1667	8.75358
		95% Confidence Interval for Mean	Lower Bound	-37.6685	
			Upper Bound	7.3351	
		5% Trimmed Mean		-14.1185	
		Median		-3.6000	
		Variance		459.751	
		Std. Deviation		21.44180	
		Minimum		-51.00	
		Maximum		1.80	
		Range		52.80	
		Interquartile Range		35.70	
		Skewness		-1.245	.845
		Kurtosis		-.020	1.741
	LOW REVENUE GROWTH	Mean		5.2200	1.25116
		95% Confidence Interval for Mean	Lower Bound	1.7462	
			Upper Bound	8.6938	
		5% Trimmed Mean		5.3556	
		Median		6.2000	
		Variance		7.827	
		Std. Deviation		2.79768	
		Minimum		.50	
		Maximum		7.50	
		Range		7.00	
		Interquartile Range		4.45	
		Skewness		-1.667	.913
		Kurtosis		2.860	2.000

Fig. 3.5 Descriptive statistics related to firms with negative and low revenue growth

of variance or normality, then statistics such as those of S-W, K-S and Levene offer more deterministic evidence.

Just as diagrams and charts such as those described in the previous chapter permit initial examination of gathered data, so too do *frequency tables* of single variables and *cross tabulations* of two or more variables. Obviously, studying

Tests of Normality

	Growthgp	Kolmogorov-Smirnov[a]			Shapiro-Wilk		
		Statistic	df	Sig.	Statistic	df	Sig.
Revenue	NEGATIVE GROWTH	.367	6	.011	.783	6	.041
	LOW REVENUE GROWTH	.269	5	.200[*]	.834	5	.150
	MEDIUM REVENUE GROWTH	.264	4	.	.951	4	.720
	HIGH GROWTH	.194	5	.200[*]	.951	5	.746

Fig. 3.6 The Shapiro–Wilks and Kolmogorov–Smirnov tests

Test of Homogeneity of Variance

		Levene Statistic	df1	df2	Sig.
Revenue	Based on Mean	8.011	3	16	.002
	Based on Median	1.654	3	16	.217
	Based on Median and with adjusted df	1.654	3	6.568	.267
	Based on trimmed mean	6.950	3	16	.003

Fig. 3.7 Results of the Levene test

variable values for hundreds of cases is unlikely to produce much understanding of the properties of the gathered data. Therefore, data are summarized or classified via diagrams and frequency tables.

If gathered data are *discrete* (e.g. integer valued), then grouping respondents into categories is straightforward. For example, respondents to a market research survey might be "highly favourable", "favourable", "indifferent", "unfavourable" or "highly unfavourable" vis-a-vis the proposed packaging of a new product. Typically, in data files, such responses could be recorded as 5, 4, 3, 2 and 1 respectively. Alternative labels for these responses would also suffice. It is a simple matter for a computer package to count the frequencies of consumer responses in these five categories and to produce a frequency table.

If the gathered data are *continuous* (e.g. weights or lengths), then the manner in which cases are grouped into categories can have a significant effect on how the characteristics of the data are interpreted. We may adopt weight categories such as "0 and under 10 g.", "10 and under 20 g.", "20 and under 30 g." etc. If, in the first category, there are many weights around 2 g. and many other weights around 8 g., then this interesting property is disguised by our grouping procedure.

The same points apply to cross tabulations of two or more variables. The manner in which the grouping of cases is performed can radically influence visual interpretation of such a frequency table. Given a cross tabulation of say two variables, it is often of interest to see if the variables are dependent upon each other. For example, as one variable (say 'student income') increases, does the other variable (say 'monthly expenditure on books') increase systematically? Thoughtless grouping procedures e.g. those that generate low frequencies can disguise an important relationship.

3.3   Univariate Frequencies

BRITISH FARMING.SAV contains estimated data that pertain to a random sample of 37 dairy farms from the population of 94 such farms. The variables in this file, along with their IBM SPSS Statistics variable names in capitals are:

* FARMNO—farm identification number,
* FARMSIZE—farm size to the nearest hectare,
* MILKGP—a grouping variable according to whether the farms annual milk fat production in 2003 was above or below the mean production. This variable takes a label of '1' if annual milk fat production is above the mean and a label of '0' if it below the mean and
* EQUIP—a variable reflecting a subjective assessment of the "status" of the farm's equipment. Labels used are '1' for high status, '2' for moderate status and '3' for low status.

These variables and their subgroups have already been labelled in BRITISH FARMING.SAV. The latter two variables above are NOMINAL. We shall initially focus on these and derive frequencies for each variable label. Having opened BRITISH FARMING.SAV, the IBM SPSS Statistics Frequencies procedure is accessed by clicking:

Analyze
  Descriptive Statistics
    Frequencies…

which generates the Frequencies dialogue box of Fig. 3.8. The variables MILKGP and EQUIP are selected from the variable list and entered into the 'Variable(s)' box via the arrow button. Clicking the Statistics… button to the right of this dialogue box in Fig. 3.8 produces the Frequencies: Statistics dialogue box of Fig. 3.9. Given such nominal data, measures like the mean, standard deviation, skewness etc. make little sense. Only the mode might be useful, but with so few grouping classes, the modal class should be obvious from the generated frequency table. Click the Continue button to return to the Frequencies dialogue box of Fig. 3.8.

Clicking the Charts… button at the bottom of the dialogue box produces the Frequencies: Charts dialogue box of Fig. 3.10. As may be seen from this figure, the generation of no charts is the default. Just for interest, select (click) Histogram(s). This option will generate histograms of MILKGP and EQUIP. There is no reason to add the normal curve to such nominal data. Click the Continue button to return to the Frequencies dialogue box. Lastly, click the Format… button to generate the Frequencies: Format dialogue box of Fig. 3.11. This dialogue box permits modification of the frequency table output. The order in which data values are sorted and displayed in frequency tables is self-explanatory.

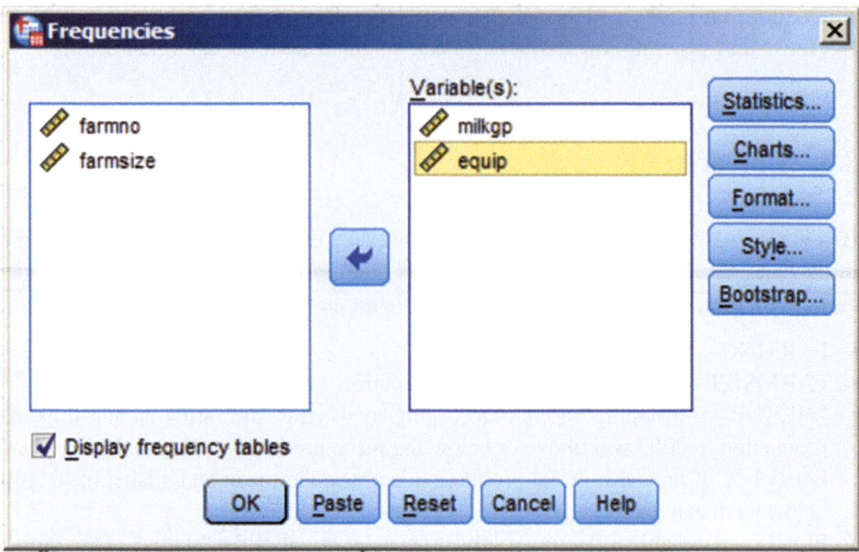

Fig. 3.8 The Frequencies dialogue box

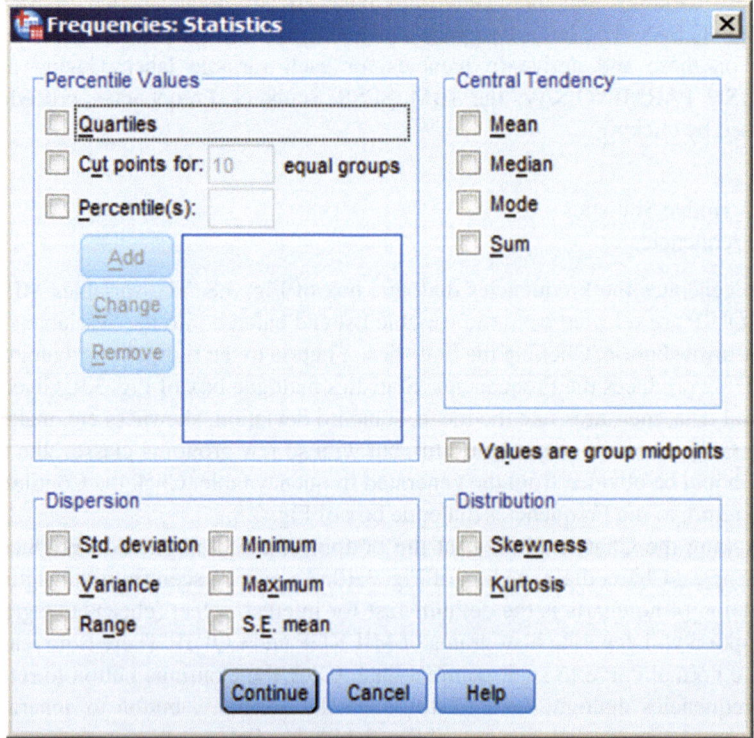

Fig. 3.9 The Frequencies: Statistics dialogue box

Fig. 3.10 The Frequencies:
Charts dialogue box

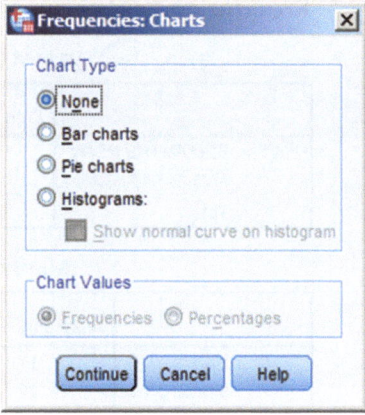

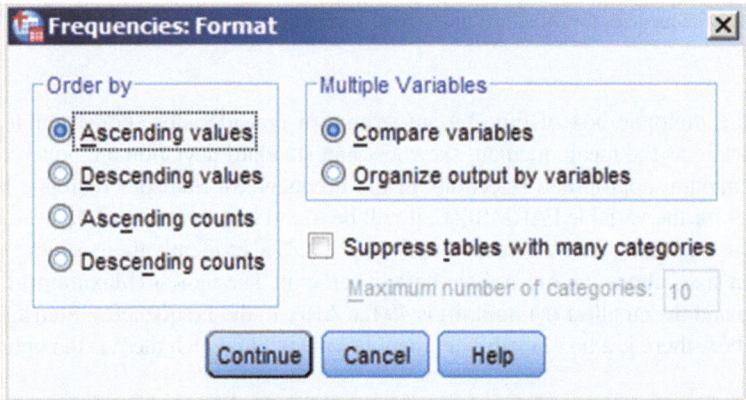

Fig. 3.11 The Frequencies: Format dialogue box

The frequency results for MILKGP and EQUIP are presented in Fig. 3.12. Although there are no missing values in this data file, such are readily accommodated in IBM SPSS Statistics as is described at the start of the next chapter. There are 37 valid cases (farms) and all are included in the frequency counts. Although there are not many groups, such frequency counts as in Fig. 3.10 are more readily comprehensible than is the result of scanning the raw data file BRITISH FARMING.SAV by eye. The low frequency of farms with low 'status of equipment' is immediately obvious, for example. Also is the fact that over 90 % of farms have equipment of at least moderate 'status'.

If we wished to examine the variable FARMSIZE, then the above IBM SPSS Statistics selections made for MILKGP and EQUIP would not be sensible. Virtually all of the farm sizes are different, so frequency tables such as those included in Fig. 3.12 would not useful, in that most frequencies would be 1. Similarly, the modal farm size would be meaningless. However, more of the statistics in the Frequencies:

Frequency Table

MILKFAT PRODUCTION GROUPING

		Frequency	Percent	Valid Percent	Cumulative Percent
Valid	BELOW THE MEAN	21	56.8	56.8	56.8
	ABOVE THE MEAN	16	43.2	43.2	100.0
	Total	37	100.0	100.0	

STATUS OF EQUIPMENT

		Frequency	Percent	Valid Percent	Cumulative Percent
Valid	HIGH	11	29.7	29.7	29.7
	MODERATE	23	62.2	62.2	91.9
	LOW	3	8.1	8.1	100.0
	Total	37	100.0	100.0	

Fig. 3.12 Frequencies for MILKGP and EQUIP

Statistics dialogue box of Fig. 3.9 are worthy of consideration. For example, such parameters as the mean, median, skewness and standard deviation are now relevant. Upon making appropriate selections in the Frequencies: Statistics dialogue box of Fig. 3.9 for the variable FARMSIZE, it will be found that the mean of this variable is just over 92.5 ha, with a skewness coefficient of 0.815. Evidently there is more spread of farm sizes above the mean than there is below it. The largest (Maximum) farm is 160 ha and the smallest (Minimum) is 49 ha. Also, in the Frequencies: Statistics dialogue box, there is a box labelled 'Percentile Values', in which there is the option:

Cut points for n equal groups.

The default, if this option is selected, is for ten approximately equally sized groups. This option would estimate the 10th, 20th, 30th, 90th percentiles for the variable at hand. The 30th percentile for the variable FARMSIZE would be an estimate of the acreage below which 30 % of the farms lie. (With 37 farms, this percentile is not an integer, so interpolation would be used by IBM SPSS Statistics). The 50th percentile is the median. The above default figure may be changed by clicking the relevant box and typing in an alternative.

3.4 Cross Tabulation of Two Variables

In this section, we are going to build up a two-variable frequency table. In particular, it would seem reasonable that the variables FARMSIZE and MILKGP are related, in that we would expect levels of milk fat production to be higher for the larger sized dairy farms, all other factors being equal. The generation of a frequency table involving (here) two variables is called a *cross tabulation* or *contingency table*. There are several statistics available in contingency tables and which examine whether there is a significant relationship between the variables at hand. The most well-known statistic is *chi square* (χ^2) and is discussed in Sect. 3.4.3. Other available statistics are described in Sect. 3.4.4.

The problem that we have in producing such a table in the present instance is that FARMSIZE involves numerical values that would cause most of the frequency counts to be one. One option is to group the values of this variable into two or more categories. We do not want this grouping to result in many frequencies that are small. Here, we will group the values of FARMSIZE according to whether they are above or below the mean of 92.5 ha. Farm sizes below the mean will be labelled as '0'; those above the mean as '1'. This grouping procedure is accomplished via the IBM SPSS Statistics Recode procedure.

3.4.1 The Recode Procedure

By means of the Recode procedure, we shall create a new variable according to whether the farm sizes are above or below the mean figure. We will have to name this new variable. Therefore, call this new variable RECSIZE, which will take labels of '0' or '1' as defined above.

To access the Recode procedure, make sure that BRITISH FARMING.SAV is open and from the Data Editor click:

Transform
  Recode
    Into Different Variables…

which produces the Recode Into Different Variables dialogue box of Fig. 3.13.

If we had selected the option:

…
    Into Same Variables…

then the values of the existing variable would be recoded and replaced for the duration of this session. In this instance, we want to create a new variable (called RECSIZE, say) and preserve the variable FARMSIZE for potential further analysis during this session.

In the dialogue box of Fig. 3.13, we click the variable name FARMSIZE into the box headed 'Numeric Variable → Output Variable'. In the box labelled 'Output Variable', click the box associated with 'Name' and type in the variable name RECSIZE. This new variable may be labelled—here as 'Recoded Farm Size'—by typing this into the box associated with Label. Click the Change button which is now active.

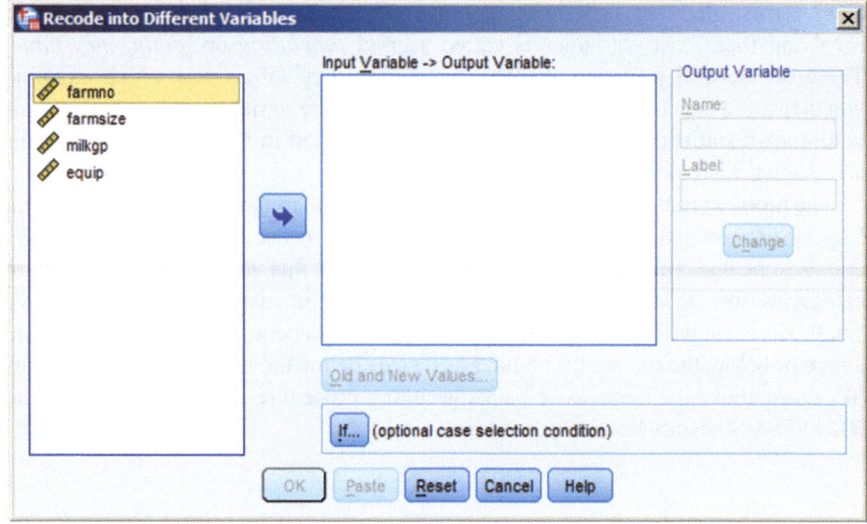

Fig. 3.13 Recode into Different Variables dialogue box

Towards the bottom of the dialogue box of Fig. 3.13 are two buttons labelled If…
and Old and New Values…The former permits conditional recoding. Here, we wish
to recode old values of FARMSIZE into new ones, so click the latter button, which
produces the dialogue box of Fig. 3.14. This dialogue box lets the user to define the
rule(s) by which the recoding is to be performed. On the left of the above dialogue
box is the option:

Range, LOWEST through value

Which will allow us to recode the farm sizes from the lowest up to the mean of
92.5 ha with the label '0'. Click inside the circle to the left of this option. Click inside
the associated box and type in 92.5. Go to the box titled 'New Value' and type in a
zero in the box associated with Value. Click the Add button. This recoding rule will
appear in the box labelled 'Old → New'. We now recode the larger farms by a similar procedure. This time click the circle associated with:

Range, value through HIGHEST

and type in 92.5 into the provided box. In the box headed New Value, type in a 1 in
the box associated with Value. Again click the Add button and this recoding rule will
also appear in the box titled 'Old → New'. The dialogue box will now appear as per
Fig. 3.14. Click the Continue button to return to the Recode into Different Variables
dialogue box of Fig. 3.13. Click the OK button and the new variable RECSIZE is
computed and added to our data file. If the user wishes to label the values of '0' and
'1' for RECSIZE (e.g. as respectively "Below the Mean" and "Above the Mean"),

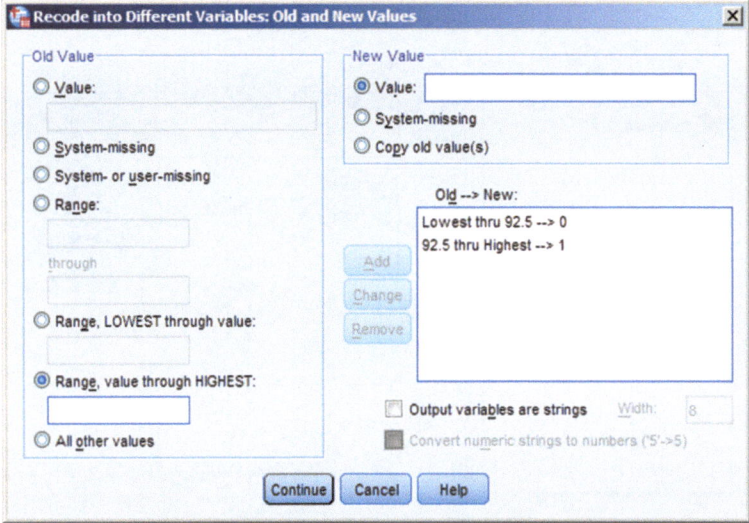

Fig. 3.14 Defining old and new values

then this is achieved by clicking from the Variable View in the IBM SPSS Statistics Data Editor and clicking the small grey box in the column titled 'Values' for the variable RECSIZE. If the data file is not saved at the end of the session, the new variable RECSIZE will be lost.

3.4.2   The IBM SPSS Statistics Crosstabs Procedure

We are now in a position to generate a contingency table involving the recoded farm sizes and the milk fat production levels. Given that RECSIZE and MILKGP each have two categories, this is referred to a 2 (rows)×2 (columns) contingency table. The Crosstabs procedure is flexible and the user has much control as to what is entered into the generated tables. For example, row and column percentages may be included. To assess the Crosstabs routine, from the Data Editor click:

Analyze
    Descriptive Statistics
        Crosstabs…

which generates the Crosstabs dialogue box of Fig. 3.15. Click the variables RECSIZE and MILKGP from the variable list and via the arrow buttons, enter one into the box headed Row(s) and the other into that headed Column(s). In this instance, it matters little which variable constitutes the rows of the table and which the columns. Clicking the Statistics… button at the bottom of the Crosstabs

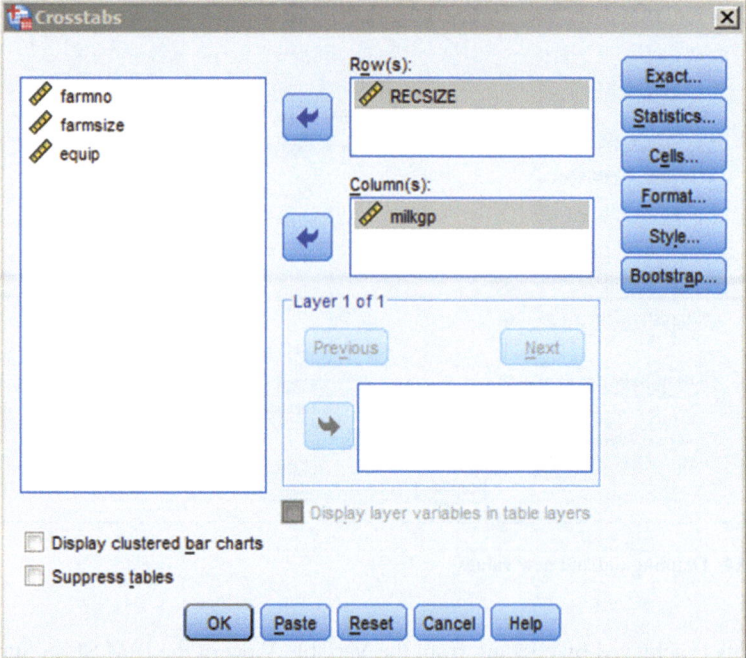

Fig. 3.15 The Crosstabs dialogue box

dialogue box produces the Crosstabs: Statistics dialogue box of Fig. 3.16. Click the box associated with chi square to select this statistic. The computation and interpretation of chi square is discussed in the next subsection of this chapter. Some of the other available statistics in Fig. 3.16 are discussed in Sect. 3.4.4. Click the Continue button to return to the Crosstabs dialogue box.

Clicking the Cells... button at the bottom of the Crosstabs dialogue box produces the Crosstabs: Cell Display dialogue box of Fig. 3.17. This latter dialogue box controls the information reported in the contingency table. Obviously, we require the Observed counts. By clicking the appropriate boxes, row, column and/or total percentages may be obtained. (Here, I have selected just Row and Column percentages). A *residual* in a contingency table is defined as the difference of the observed cell count and the expected cell count that would be obtained if the variables in the problem are independent. Here, I have selected 'Unstandardized residuals'. These residuals may be standardized to have a mean of zero and variance of unity if desired. Click the Continue button to return to the Crosstabs dialogue box.

Clicking the Format... button produces the Crosstabs: Format dialogue box of Fig. 3.18. This enables the user to define how a table is displayed. The default is that both the variable and its value labels are printed and the row variable values are given in ascending order from lowest to highest. Click the Continue button to return to the Crosstabs dialogue box and now click the OK button to produce the required contingency table. This crosstabulation is presented in Fig. 3.19.

Fig. 3.16 The Crosstabs:
Statistics dialogue box

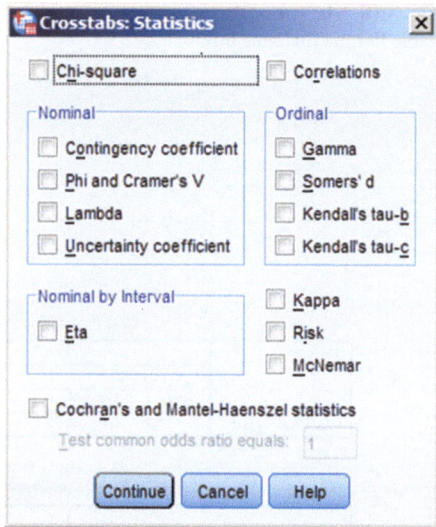

Fig. 3.17 The Crosstabs:
Cell Display dialogue box

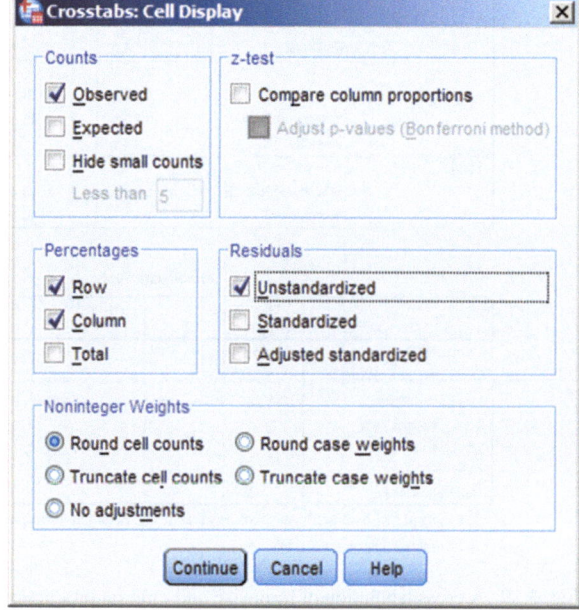

In the contingency table, the count is followed by the row percentage, then the
column percentage and then the unstandardized residuals. For example, there are 18
farms with acreages below the mean and with milkfat production levels below the
mean. This frequency represents 78.3 % of the 23 farms in the first row and 85.7 %
of the 21 farms in the first column. Turning to the margins of the table, there are 14
farms above the mean size, which represents 37.8 % of the sample of 37 farms.

Fig. 3.18 The Crosstabs:
Table Format dialogue box

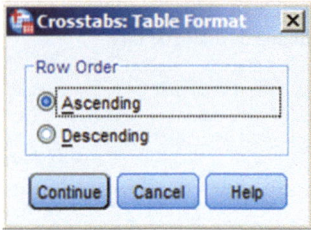

Case Processing Summary

	Cases					
	Valid		Missing		Total	
	N	Percent	N	Percent	N	Percent
RECSIZE * milkgp	37	100.0%	0	0.0%	37	100.0%

RECSIZE * milkgp Crosstabulation

			milkgp		Total
			BELOW THE MEAN	ABOVE THE MEAN	
RECSIZE	.00	Count	18	5	23
		% within RECSIZE	78.3%	21.7%	100.0%
		% within milkgp	85.7%	31.3%	62.2%
		Residual	4.9	-4.9	
	1.00	Count	3	11	14
		% within RECSIZE	21.4%	78.6%	100.0%
		% within milkgp	14.3%	68.8%	37.8%
		Residual	-4.9	4.9	
Total		Count	21	16	37
		% within RECSIZE	56.8%	43.2%	100.0%
		% within milkgp	100.0%	100.0%	100.0%

Chi-Square Tests

	Value	df	Asymp. Sig. (2-sided)	Exact Sig. (2-sided)	Exact Sig. (1-sided)
Pearson Chi-Square	11.453[a]	1	.001		
Continuity Correction[b]	9.254	1	.002		
Likelihood Ratio	11.982	1	.001		
Fisher's Exact Test				.002	.001
Linear-by-Linear Association	11.143	1	.001		
N of Valid Cases	37				

a. 0 cells (0.0%) have expected count less than 5. The minimum expected count is 6.05.

b. Computed only for a 2x2 table

Fig. 3.19 A cross tabulation of farm size and milk fat production

Figure 3.19 does suggest that milkfat production levels increase with farm size, as would be expected. Only eight farms run counter to this conclusion. A less subjective examination of this contention is available through the statistics that are presented along with the cross tabulation.

3.4.3 Calculation and Interpretation of the Chi Square Statistic

The chi square statistic tests whether the row and column variables in a contingency table are independent. For example, if they are indeed independent, the probability that a farm is below the mean for RECSIZE and below the mean for MILKGP is:

$$P(\text{Below mean RECSIZE}) * P(\text{Below mean MILKGP})$$

This product of the individual probabilities is derived from the multiplication law of probability and under the assumption that RECSIZE and MILKGP are statistically independent. With reference to the marginal totals in Fig. 3.19, we may compute that:

$$P(\text{Below mean RECSIZE and below mean MILKGP}) = (21/37)*(23/37)$$
$$= 0.3528.$$

Assuming independence, the expected number of farms with values of these two variables below their respective means is, therefore, $(0.3528)*37 = 13.05$, compared with our observed number of 18 such farms. Denoting the marginal frequency of column 1 as C (here, 21) and that of row 1 as R (here, 23), it will be noted that in the above calculation of the expected frequency, a 37 cancels (i.e. $(21*23)/37 = 13.05$), so a simplified way of computing the expected value in this cell would thus be:

$$R * C / N,$$

where N is the total number of readings in the table. The difference between the observed frequency of 18 and the expected frequency of 13.05 is +4.95. This is the residual reported (to one decimal place) in Fig. 3.19. If the sizes of farms and their milk fat levels were independent, we would expect less farms to fall into this category than we have actually observed. The expected values for the remaining cells in the table can be computed in a similar manner.

Such computations are based on the assumption that RECSIZE and MILKGP are statistically independent. This is, in fact, the central idea upon which computation of the chi square statistic is to be based. The key point is that if a belief that the two study variables are independent is reasonable, then the expected frequencies (E) should be consistently close to those observed (O). This assertion is examined by computing:

$$\chi^2 = \sum \frac{(O-E)^2}{E}$$

Which is distributed as the chi square statistic and where O and E are as previously defined. If the independence assumption is tenable, the E and O will be numerically close and this total will be small because the numerator of the above equation will be close to zero. The larger is the total, the less likely is the independence assumption to be true. The chi square statistic is reported towards the bottom of Fig. 3.19. IBM SPSS Statistics reports a numerical value for the significance associated with the computed value of the chi square statistic. In the case of a contingency table, the significance is the probability of obtaining a chi square statistic as extreme as the one computed under the assumption that the variables at hand are independent. As before, low probabilities (associated with relatively large chi square values) of conventionally below 0.05 would lead us to doubt the validity of the independence assumption.

In the case of 2×2 tables such as in the present example, it is advocated that the formula for chi square should be modified by the inclusion of what is called Yates' continuity correction. In this instance **ONLY**, the user should refer to the row labelled 'Continuity correction' in the output Fig. 3.19. For all other sized contingency tables (i.e. not 2×2), the row titled 'Pearson chi square' must be used.

A further point about the chi square contingency statistic is that it should not be used if the expected frequencies are small. Specifically, one should not use the chi square contingency statistic if (i) any cell has an expected value less than unity or (ii) more than 20 % of the cells have expected values less than five in contingency tables larger than 2×2. If any expected frequency in a 2×2 contingency table is less than five, then IBM SPSS Statistics automatically uses what is called Fisher's exact test instead of chi square to assess the notion of independence and an associated level of significance is naturally reported. In the present example, we do have a 2×2 contingency table. The numerical value of chi square with the Yates' continuity correction is 9.254. Crucially, the significance of 0.002 associated with this statistic suggests that there is here a very low probability of obtaining a chi-square value as extreme as 9.254 if the variables are independent. We thus doubt the independence assumption and conclude that milk fat production levels do depend on farm size.

The unstandardized residuals of Fig. 3.19 show the direction of the relationship. For example, we observe 11 farms that are of above average size and which are producing above average milkfat levels. The (unstandardized) residual of 4.9 means that we observe 4.9 more farms in this cell of the contingency table than we would have expected if farm sizes and production levels were independent. As concluded earlier, larger farms generally have larger milkfat production levels. The variables are not independent.

The Mantel-Haenszel statistic is reported in Fig. 3.19 and is also distributed as chi square. It is found in the row labelled 'Linear-by-Linear Association'. It tests whether the variables under study are linearly related. Given that our labels of '0' and '1' for RECSIZE and MILKGP are nominal, this latter test possesses little meaning. The likelihood ratio is again distributed as chi square and is based on what is called 'maximum likelihood theory' in Statistics. Suffice it to say, it is an alternative to the Pearsonian chi square and for large samples, the two statistics yield very similar results. The minimum expected frequency is reported by default by IBM SPSS Statistics to assess whether chi square should be used as per the rules cited earlier for tables greater than 2×2. (Remember that in the 2×2 case such as in our example, IBM SPSS Statistics would have automatically used Fisher's exact test instead of chi square to assess independence should any expected frequency have been less than five). If more than 20 % of the cells in a table have expected frequencies of less than five, then IBM SPSS Statistics will be report this fact.

3.4.4 Other Statistics Available in the Crosstabs Procedure

The Crosstabs: Statistics dialogue box of Fig. 3.16 presents several measures of association, which attempt to quantify the relationship between variables in cross tabulations involving nominal and ordinal data. These measures vary in their interpretation and no measure is the optimal in all situations. The properties of the various measures should be considered when deciding which to implement. The measures of association for nominal data are briefly discussed below.

The phi statistic may lie between 1 and -1 inclusive, indicative of positive and negative association.

Cramer's V is also a measure of association based on chi square and lies between 0 (variables are independent) and 1 (variables are dependent).

Pearson's *contingency coefficient* becomes zero when the variables are not associated. The upper limit depends on the number of rows and columns in the contingency table. In the 2×2 case, the upper limit is $\sqrt{0.5} = 0.707$. *Lambda* (Λ) is called a proportional reduction in error (PRE) measure of association, in that it reflects the reduction in error when the values one variable, are used to predict the values of the other variable. If $\Lambda = 1$, the independent variable perfectly predicts the dependent variable and if $\Lambda = 0$, the independent variable is of no use in predicting the dependent variable.

The *uncertainty coefficient* is another PRE measure. The closer it is to unity, the more information is provided about the value of the second variable from knowledge of the value of the first variable. It has a lower bound of zero, when no information about the value of the second variable is obtained from knowledge of the first. In the *Crosstabs: Statistics dialogue box* of Fig. 3.16, there are measures of association used when order is present in the data, for example, Somer's d and gamma. These also assess the strength of association between two variables.

3.5 Customizing Tables

It is possible for the user to customize tabular output for presentation purposes. To illustrate this option, we shall examine the relationship between levels of milkfat production (MILKGP) and the "status" of farm equipment (EQUIP) over the two groups of farm sizes (RECSIZE). IBM SPSS Statistics will generate a table of MILKGP against EQUIP for farm sizes below the mean of 92.5 ha and a second table of MILKGP against EQUIP for farm sizes above 92.5 ha. The user has much control over what is reported in such tables as well as their physical format. The data need to be nominal or ordinal. Therefore, go to the Variable View in the IBM SPSS Statistics Data Editor and in the column headed 'Measure' change RECSIZE from scale (ratio or interval) measurement to nominal measurement. To customize a table, from the Data Editor click:

Analyze
    Tables
        Custom Tables…

which produces the Custom Tables dialogue box of Fig. 3.20. The table will be layered according to the values of '0' and '1' applied to RECSIZE. This will result in two separate tables of MILKGP against EQUIP being generated, one for each of these two numerical values.

In these tables, MILKGP is chosen to be the row variable, EQUIP the column variable and RECSIZE as the layer variable. Click the variable name MILKGP from the list and drag it into the (vertical) Rows box; similarly drag EQUIP into the (horizontal) columns box. Drag RECSIZE into the 'Layer' box'.

Click the 'Test Statistics' tab, which generates the dialogue box of Fig. 3.21. This enables the user to perform separate chi-square tests for RECSIZE=0 and for RECSIZE=1. I have checked the box in the middle of Fig. 3.21 to achieve this. Clicking the OK button generates the output in Fig. 3.22.

Note that for the farms of below mean size, the variables MILKGP and EQUIP are not independent, since the obtained significance level of 0.056 is greater than 0.05.

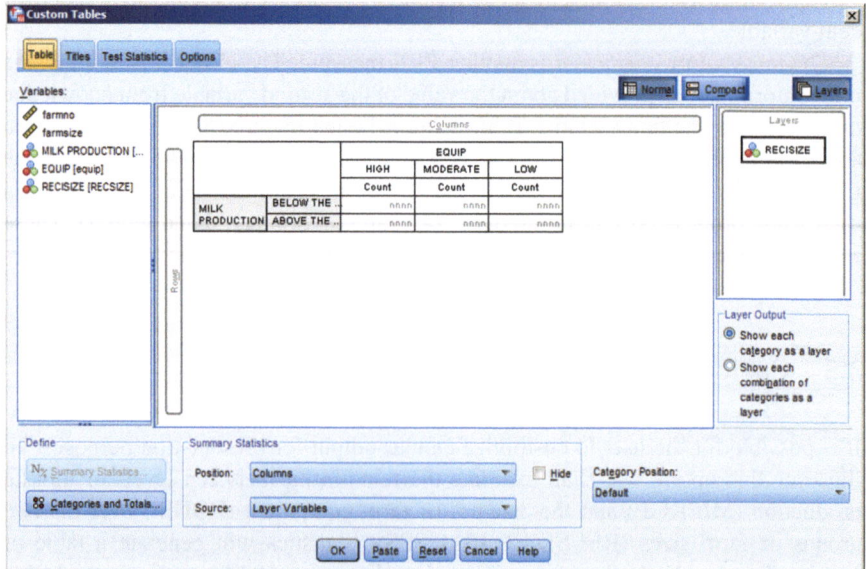

Fig. 3.20 The Custom Tables dialogue box

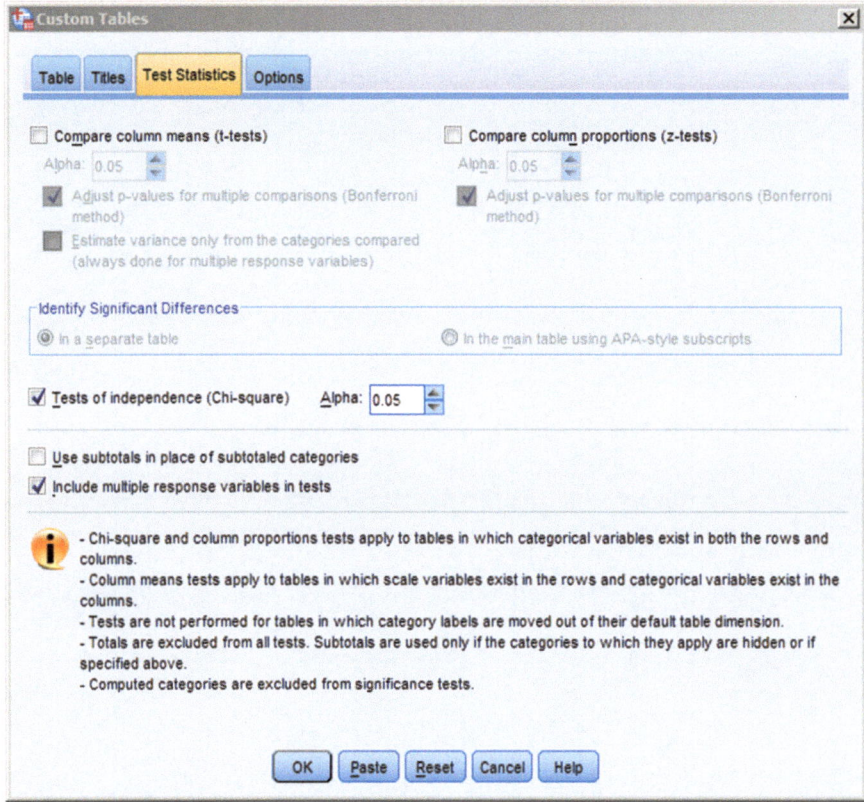

Fig. 3.21 Selecting separate chi-square tests

Fig. 3.22 Output from customizing tables

Custom Tables

RECISIZE .00

		EQUIP		
		HIGH	MODERATE	LOW
		Count	Count	Count
MILK PRODUCTION	BELOW THE MEAN	2	13	3
	ABOVE THE MEAN	3	2	0

Pearson Chi-Square Tests

RECISIZE .00

		EQUIP
MILK PRODUCTION	Chi-square	5.759
	df	2
	Sig.	.056[a,b]

Results are based on nonempty rows and columns in each innermost subtable.

a. More than 20% of cells in this subtable have expected cell counts less than 5. Chi-square results may be invalid.

b. The minimum expected cell count in this subtable is less than one. Chi-square results may be invalid.

Chapter 4
Coding, Missing Values, Conditional and Arithmetic Operations

As we have already seen, some data such as that gathered in market research, social and economic surveys etc., are not numerical e.g. respondents' sex, their levels of education or voting intentions. These data are called nominal as discussed in Sect. 1.2.2. In data files, one way such items are entered is to assign numbers or symbols to the various responses. This allocation of labels is known as *coding the data* and is discussed in the first section of this chapter.

Also, in some statistical routines, it may be necessary mathematically to transform the data to meet the particular assumptions of statistical techniques used. Typically, this involves the computation of new variables. Generally, operations whereby one takes existing variables and alters their values or uses them to create new variables are called *data transformations*. Such transforms may be *numeric* e.g. to convert from one currency to another we would have to compute a new variable. Some transforms may be *conditional*. For example, we may wish to compute VAT rates for a subset of the products in our data file. These form the second and third sections of this chapter. Having created new variables by means of data transformations, we may wish to list cases in rank order in terms of the new variable. This may be achieved by the auto recode procedure of the last section of this chapter.

4.1 Coding of Data

Coding schemes are arbitrary by their very nature. A type of drink regularly consumed could be coded as L for light, B for bitter, M for mild and O for all other types. The gender of respondents to market research surveys is often coded as 0 for female and 1 for male (or vice versa). Consumers' attitudes towards a new product could be coded as 5 for "highly satisfied", 4 for "satisfied" 3 for "neither", 2 for "unsatisfied" and 1 for "very unsatisfied". Groups of consumers regarded as "high", "medium" and "low" levels of credit risk could be coded as 1, 2 and 3 respectively. It is common to reserve codes for responses such as "don't know" and "unwilling to answer".

© Springer International Publishing Switzerland 2016
A. Aljandali, *Quantitative Analysis and IBM® SPSS® Statistics,*
Statistics and Econometrics for Finance, DOI 10.1007/978-3-319-45528-0_4

Generally, whether data are gathered from official sources or questionnaires, there should be a code(s) reserved for missing values. IBM SPSS Statistics has a special facility for dealing with missing observations e.g. one option is to omit such cases from all calculations.

4.1.1   Defining Missing Values

In this example the data pertain to levels of library provision in 17 London boroughs. The file LIBRARY.SAV contains the following variables:

- BOROUGH—the name of each borough: a string variable,
- BOOKEXP—expenditure (£) per 1000 inhabitants on books,
- PAPEREXP—expenditure (£) per 1000 inhabitants on newspapers,
- TOTEXP—total library expenditure (£) per 1000 inhabitants,
- STAFF—the number of library staff employed and
- POPN—the population of each borough in thousands of persons.

The Variable View accessible from the IBM SPSS Statistics Data Editor permits the definition of any missing value(s). Missing values should be given discrete codes such as 9, 99 or 999—in fact anything that will not confuse. In a market research survey question involving "yes" (code 1) versus "no" (code 0) responses, we may reserve a code of 9 for refusal to respond. We could also reserve a code of 8 for "don't know". In the file LIBRARY.SAV, note that there is a reading of 9 for the variable TOTEXP for the BOROUGH of Kingston. This is a missing observation. I have not declared this as a missing value in LIBRARY.SAV, so it is necessary to do so.

Figure 4.1 presents the Variable View for the IBM SPSS Statistics data file LIBRARY. SAV. In the row for the variable TOTEXP and in the column labelled 'Missing', note that there are currently no missing values declared. Click the word None in this cell and then the small grey box, which produces the Missing Values dialogue box of Fig. 4.2. The user may define three, discrete missing value codes. Alternatively, the user may define a range of data values as missing, plus one discrete missing value.

Click the 'Discrete missing values' option in Fig. 4.2 and enter 9 in one of the available boxes. Click the OK button and the Variable View will now appear as per Fig. 4.3. If the user had not defined the code of 9 as missing, then this value would have been used in statistical computations leading to strange results e.g. derivation for the mean value of TOTEXP.

4.1.2   Types of Missing Value

In the above example, the code of 9 ascribed to TOTEXP for the BOROUGH of Kingston is called a *user-defined missing value*. IBM SPSS statistical procedures and data transformations recognise this flag and any case with user-defined missing

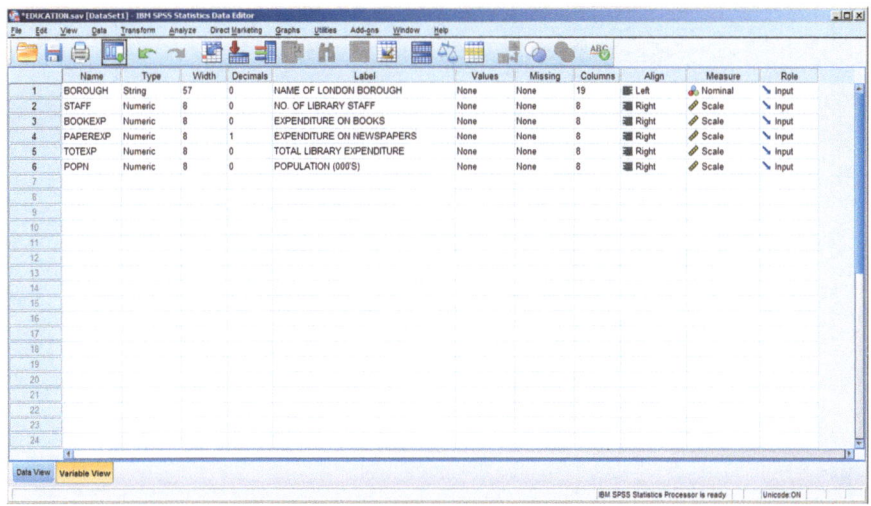

Fig. 4.1 The Variable View for the file LIBRARY.SAV

Fig. 4.2 The Missing
Values dialogue box

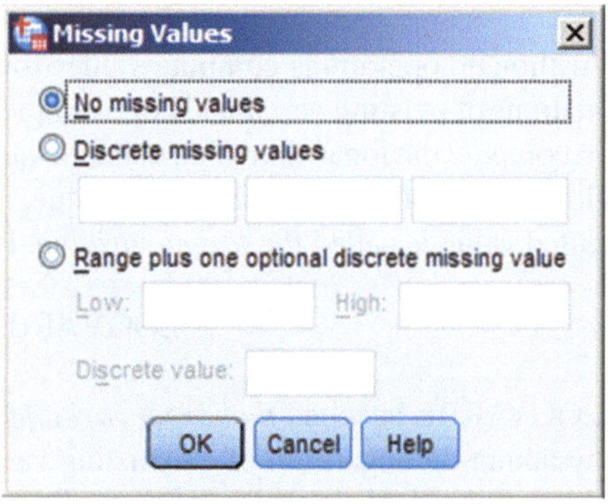

values, is handled specially. For example, in frequency tables a code of 9 could represent a missing value because of "failure to respond". The frequency with which the code of 9 occurred will be counted in the frequency table. However and as mentioned previously, due to its being declared as missing this code is not allowed to contribute to, for example, cumulative frequency calculations. Any blank numeric cells in a data file are assigned *a system-missing value*, which is indicated with a full stop (.). This situation could occur, for example, if the user wished to compute a new variable Z (see Sect. 4.2) defined as $Z = X/Y$, where X and Y were existing variables, but Y had a value of zero for a case(s).

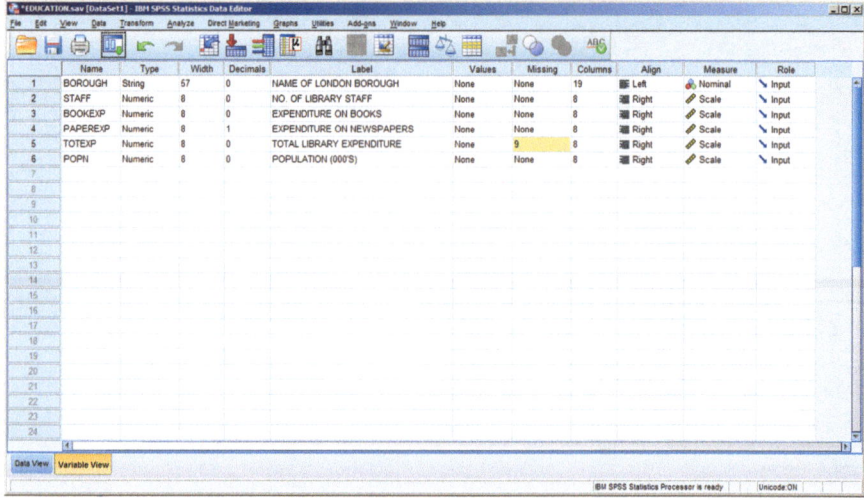

Fig. 4.3 The Variable View with a declared missing value

4.2   Arithmetic Operations

Arithmetic operations compute values for a variable, based on numerical transformations of existing variables. For example, in forecasting, it is sometimes necessary to compute the logarithm of an existing variable X to meet certain statistical assumptions underlying a particular technique. The new variable that receives the computed value is called the *target variable*. For example, in:

$$LOGVARBL = LN(X),$$

LOGVARBL is called *the target variable* and is computed by IBM SPSS Statistics by taking the logarithm of an existing variable, X. In our present example, we will compute book plus newspaper expenditure for each borough as a percentage of their total library expenditure. It will be recalled that all expenditures are measured in £ per 1000 inhabitants. We shall call the target variable which receives these percentages, RATIO. Arithmetically, this may be stated as:

$$RATIO = ((BOOKEXP + PAPEREXP) / TOTEXP) * 100.$$

The slash (/) represents division and the star (*) represents multiplication. The only other of the available operators that is possibly not obvious is that ** represents exponentiation, or "rising to a power". It is absolutely vital to consider the order in which operations are performed. Functions are evaluated first, then * and / and finally + and −. Brackets are of great use in the control of operations.

   To perform the above computation of RATIO for each borough, from the Data Editor, click:

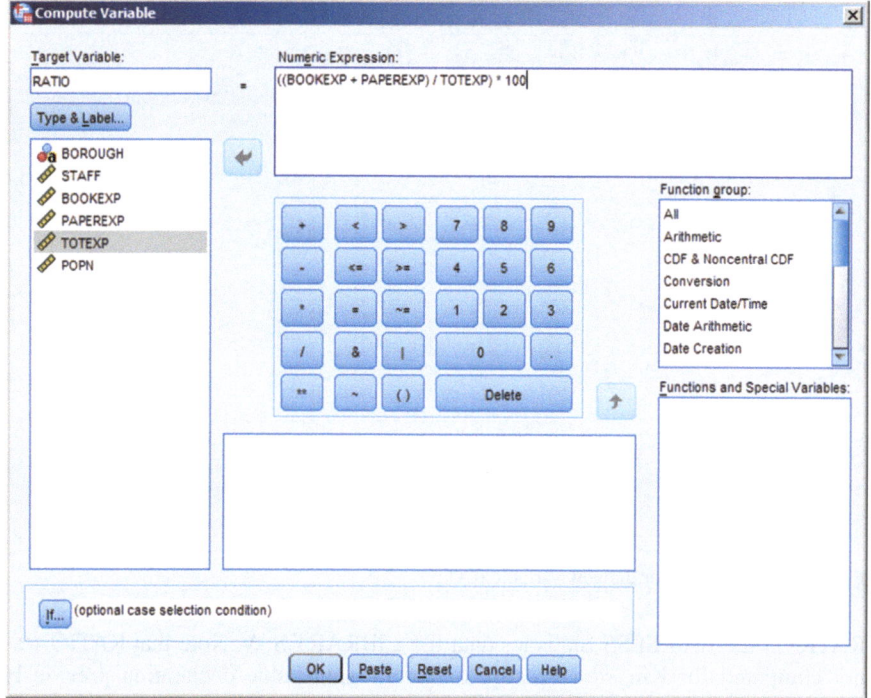

Fig. 4.4 The Compute Variable dialogue box

Transform

Compute Variable…

which gives rise to the Compute Variable dialogue box of Fig. 4.4.

Click the box below the heading 'Target Variable' to enter the name of this new variable, i.e. RATIO, as shown in Fig. 4.4. By clicking the Type & Label… button, we may define RATIO as numeric and a label may be applied to this variable, say "Expenditure Ratio", omitting the speech marks. Under the heading 'Numeric Expression', we enter the transformation. Click this box and type in the beginning of the mathematical expression of the equation on the previous page i.e. the two brackets ((Click the variable name BOOKEXP and enter it into the expression via the arrow button. The + sign may be typed directly or selected (clicked) from the calculator pad in the Compute Variable dialogue box. Next click the variable PAPEREXP and enter it into the expression via the arrow button. Complete the remainder of the expression of the equation in a similar fashion. To operationalize, click the OK button at the bottom of the dialogue box.

The variable RATIO is now computed for each BOROUGH and the numerical values of this variable are added in the next available column at the right of LIBRARY. SAV. This file may be saved under the name LIBRARY.SAV (i.e. it replaces the old version) or under a new name. Figure 4.5 shows the computation of the variable

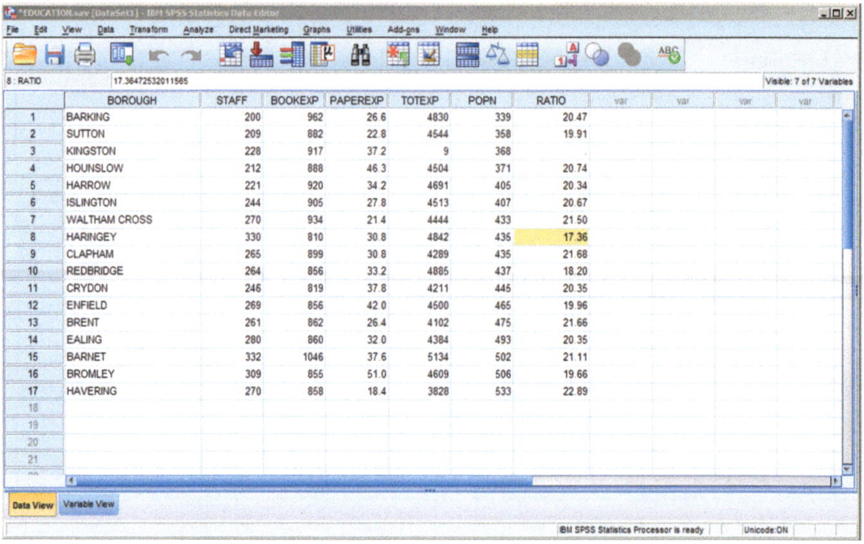

	BOROUGH	STAFF	BOOKEXP	PAPEREXP	TOTEXP	POPN	RATIO
1	BARKING	200	962	26.6	4830	339	20.47
2	SUTTON	209	882	22.8	4544	358	19.91
3	KINGSTON	228	917	37.2	9	368	.
4	HOUNSLOW	212	888	46.3	4504	371	20.74
5	HARROW	221	920	34.2	4691	405	20.34
6	ISLINGTON	244	905	27.8	4513	407	20.67
7	WALTHAM CROSS	270	934	21.4	4444	433	21.50
8	HARINGEY	330	810	30.8	4842	435	17.36
9	CLAPHAM	265	899	30.8	4289	435	21.68
10	REDBRIDGE	264	856	33.2	4885	437	18.20
11	CRYDON	246	819	37.8	4211	445	20.35
12	ENFIELD	269	856	42.0	4500	465	19.96
13	BRENT	261	862	26.4	4102	475	21.66
14	EALING	280	860	32.0	4384	493	20.35
15	BARNET	332	1046	37.6	5134	502	21.11
16	BROMLEY	309	855	51.0	4609	506	19.66
17	HAVERING	270	858	18.4	3828	533	22.89

Fig. 4.5 Computation of the new variable RATIO

RATIO in the IBM SPSS Statistics data file LIBRARY.SAV. Note that RATIO was not computed for Kingston because of the missing value declaration previously made. This is indicated by the full stop in LIBRARY.SAV in the relevant cell.

Via the IBM SPSS Statistics Descriptives procedure of Sect. 1.2.2, the mean ratio of book plus newspaper expenditures in relation to overall library expenditure may be shown to be nearly 20 % for these boroughs. The borough of Havering had the highest ratio of 22.89 % and Haringey the lowest at 17.36 %. There is not much variation in these ratio figures; they are pretty consistent across the boroughs examined.

It might be of interest to compute how many people in each borough are served by one member of library staff. This can be achieved by computing a new variable, say POPSTAFF, defined as:

$$POPSTAFF = POPN / STAFF,$$

remembering that the variable POPN is quoted as thousands of persons. This time, POPSTAFF will be computed for the borough of Kingston as well, because the declaration of the missing value for total expenditure does not effect this calculation. Use of the Compute procedure described above will generate numerical values for POPSTAFF.

Use of the Descriptives routine shows that this variable is remarkably consistent across the boroughs, with a mean value of nearly 1.6 (thousands) of people. Havering had the highest number of people served by one library staff member (POPSTAFF = 1.97) and Haringey the least number (POPSTAFF = 1.32). Although the borough of Haringey spent the lowest proportion of its total library expenditure of books and newspapers, it has reasonable levels of staffing provision to serve the population in a relative sense.

4.3   Conditional Transforms

Conditional expressions are used to apply transformations to a selected subset of the cases in a data file. If the result of a conditional expression is true for a particular case, then the transform required is applied to that case. If the result of the conditional expression is false or missing, the transform is not applied to that case. It may be recalled that TOTEXP is overall library expenditure in £ per 1000 people and that POPN is the population of each borough in thousands. If we multiply TOTEXP and POPN, we will derive the actual overall expenditure in £. We shall, therefore, define:

$$ACTEXP = TOTEXP * POPN.$$

However, let us introduce the notion of a conditional computation by only computing ACTEXP for the largest boroughs, say those boroughs with a population above the mean figure of 435.71 (thousands).

It will be recalled that the *Compute Variable dialogue box* presented in Fig. 4.4 is accessed by clicking:

Transform
  Compute Variable…

The 'Target variable' is ACTEXP, so type this into the pertinent box. In the box headed 'Numeric Expression' we need to insert TOTEXP * POPN. Click the variable names from the list and enter them via the arrow key. The multiplication symbol, *, may be typed in or clicked from the calculator pad. Specification of conditional expressions is achieved by clicking the If… button in this dialogue box. This produces the *Compute Variable: If Cases dialogue box* of Fig. 4.6. We wish to compute ACTEXP only for those boroughs with populations in excess of 435.71 (thousands). Therefore, in the dialogue box of Fig. 4.6, select (click) the option:

'Include if case satisfies condition'

From the variable list, select (click) the variable POPN and via the arrow key, enter it into the box under the above option. Either type in the '>' symbol or click it from the calculator pad. Either type in 435.71 or select the digits and decimal point from the calculator pad. This defines our condition that:

$$POPN > 435.71$$

Click the Continue button to return to the *Compute Variable dialogue box* and now click OK to operationalize this conditional computation. Examination of the data file LIBRARY.SAV as per the output in Fig. 4.7 will indicate that the variable

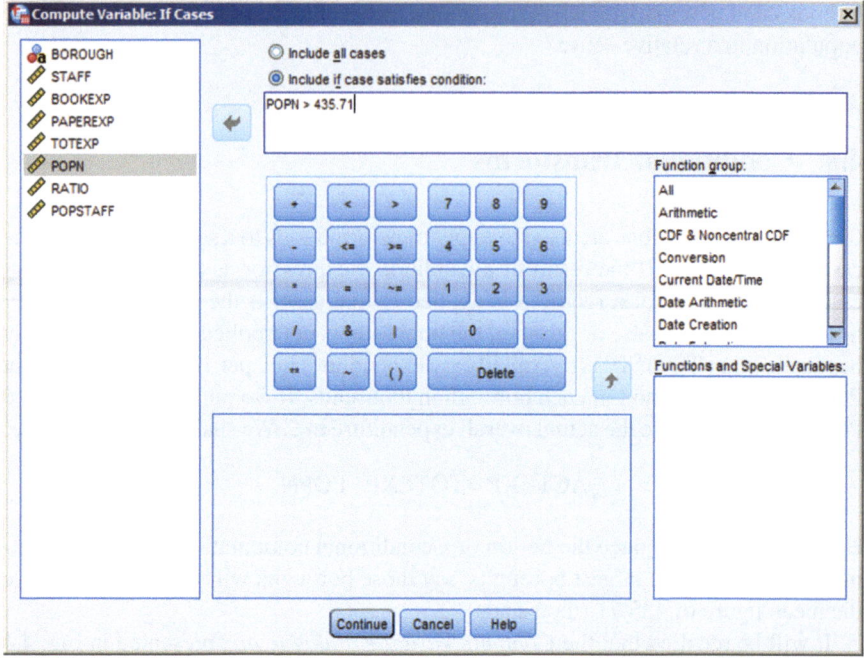

Fig. 4.6 The Compute Variable: If Cases dialogue box

ACTEXP has been computed and added to the right of the existing data. ACTEXP has been calculated only for the eight boroughs that have populations greater than 435.71. The remaining cases that fail to meet this condition have been allocated a system missing value, as evidenced by the full stop for these latter cases.

Another use of a conditional computation is to define cases that meet a particular criterion. For example, we might wish to identify those boroughs whose expenditure on books is greater than £1000 per 1000 inhabitants and whose expenditure on newspapers is greater than or equal to £25 per 1000 inhabitants. In symbols, this condition is:

$$BOOKEXP > 1000 \, AND \, PAPEREXP >= 25.$$

We shall compute a variable called HISPEND which will take a value of 1 if the above condition is met. Otherwise, a system missing value will be allocated to HISPEND.

In the *Compute Variable dialogue box*, HISPEND acts as the 'Target Variable'. In the box titled 'Numeric Expression', simply type in a '1'. Therefore, IBM SPSS Statistics will calculate HISPEND = 1, *if* the above condition is met. Click the If…

	BOROUGH	STAFF	BOOKEXP	PAPEREXP	TOTEXP	POPN	RATIO	POPSTAFF	ACTEXP
1	BARKING	200	962	26.6	4830	339	20.47	1.70	.
2	SUTTON	209	882	22.8	4544	358	19.91	1.71	.
3	KINGSTON	228	917	37.2	9	368		1.61	.
4	HOUNSLOW	212	888	46.3	4504	371	20.74	1.75	.
5	HARROW	221	920	34.2	4691	405	20.34	1.83	.
6	ISLINGTON	244	905	27.8	4513	407	20.67	1.67	.
7	WALTHAM CROSS	270	934	21.4	4444	433	21.50	1.60	.
8	HARINGEY	330	810	30.8	4842	435	17.36	1.32	.
9	CLAPHAM	265	899	30.8	4289	435	21.68	1.64	.
10	REDBRIDGE	264	856	33.2	4885	437	18.20	1.66	2134745
11	CRYDON	246	819	37.8	4211	445	20.35	1.81	1873895
12	ENFIELD	269	856	42.0	4500	465	19.96	1.73	2092500
13	BRENT	261	862	26.4	4102	475	21.66	1.82	1948450
14	EALING	280	860	32.0	4384	493	20.35	1.76	2161312
15	BARNET	332	1046	37.6	5134	502	21.11	1.51	2577268
16	BROMLEY	309	855	51.0	4609	506	19.66	1.64	2332154
17	HAVERING	270	858	18.4	3828	533	22.89	1.97	2040324

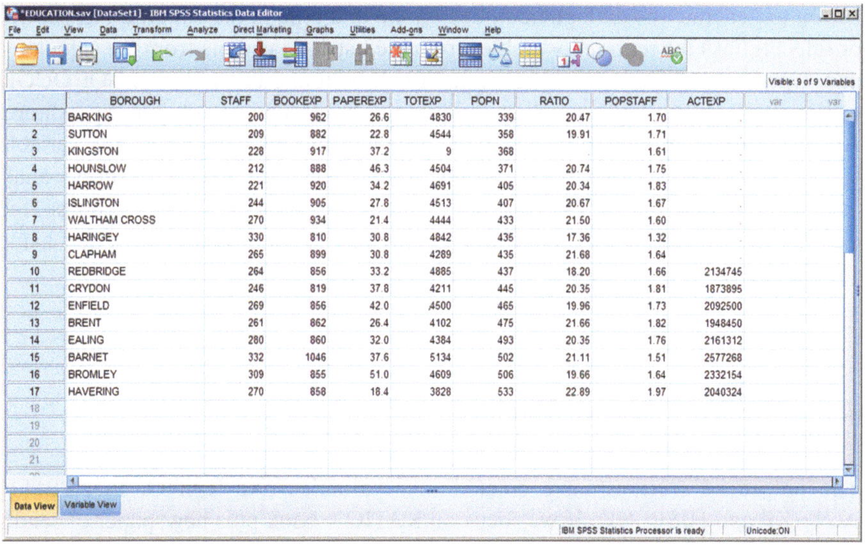

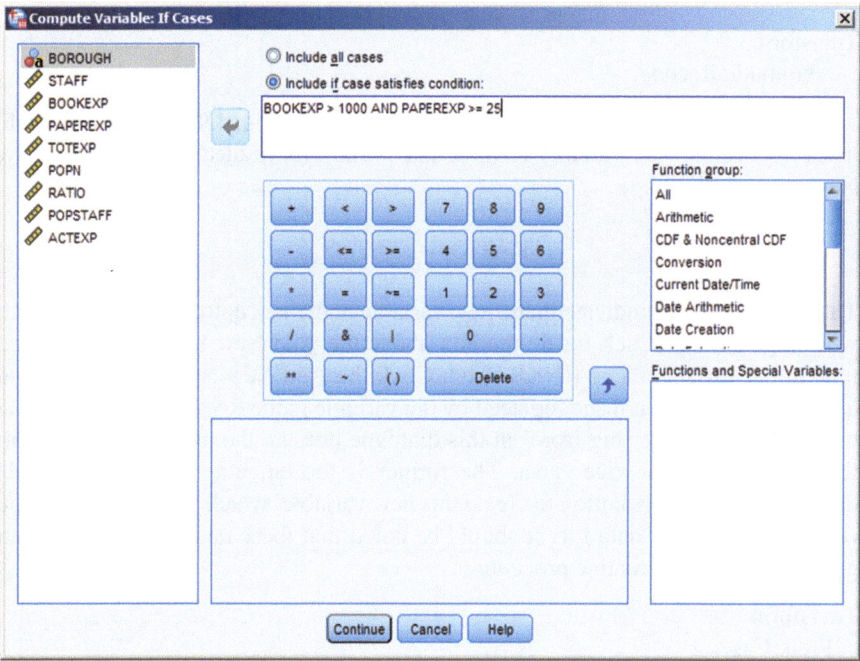

Fig. 4.7 Results of performing a conditional calculation

Fig. 4.8 Creation of the variable HISPEND

button to generate the dialogue box of Fig. 4.8 and type in the condition as shown, or select (click) the required variables from the list and symbols from the calculator pad. Upon operationalizing this routine, it will be found that only the BOROUGH of Barnet meets our high spending condition of books and newspaper expenditure. All the remaining boroughs have the system missing value for the variable HISPEND. If the user wishes to keep any newly created variable(s) for further analysis, then the data file should be saved. If the name LIBRARY.SAV is chosen, the existing file will be overwritten. If this overwriting is not desired, then a new name should be used.

4.4   The Auto Recode Facility

The preceding sections have created new variables like RATIO, POPSTAFF, ACTEXP and HISPEND. If we focus on RATIO—book plus newspaper expenditure as a percentage of total library expenditure—we may wish to rank the boroughs, say from lowest to highest on this variable. A new variable will be created and which contains these ranks. We could call this new variable RNKRATIO. We thus have recourse to the IBM SPSS Statistics Auto recode procedure, which is accessed by clicking from the Data Editor:

Transform
    Automatic Recode…

which produces the *Automatic Recode dialogue box* of Fig. 4.9. From the variable list, select (click) the variable RATIO and in the box headed 'Variable → New Name', you will see:

RATIO → ????????

The question marks indicate that a new variable name is required for the ranks that will be generated. Click inside the box above the Add New Name… button and type in the variable name RNKRATIO. Click the Add New Name button and the question marks will now be replaced by the variable name RNKRATIO. Under the heading 'Recode Starting from' in this dialogue box are the options to rank from lowest to highest or vice-versa. The former is the default. Click whichever is desired. Click the OK button to create this new variable, which will be added to the right of the existing data file. It should be noted that these ranks could have been generated by an alternative procedure:

Transform
    Rank Cases…

Fig. 4.9 The Automatic
Recode dialogue box

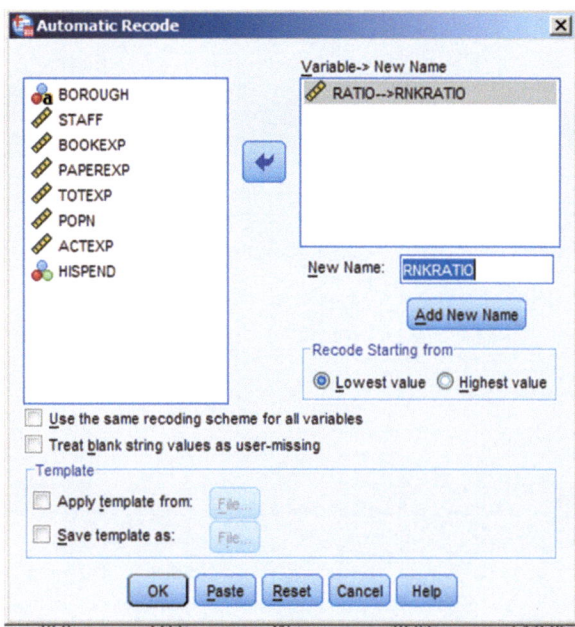

Part III
Hypothesis Tests

Chapter 5
Hypothesis Tests Concerning Means

A frequent use of statistical analysis is to make *inferences* about a population characteristic on the basis of sample evidence. Such a population characteristic could be the mean level of household expenditure on domestic insurance in a company's sales area. Rather than estimating such population characteristics per se, it is part of the in-built philosophy of IBM SPSS Statistics to test hypotheses about them. Such a process is called *hypothesis testing*. This chapter concerns hypothesis tests for population mean(s).

Many of the classical hypothesis tests employed to examine hypotheses about means are referred to as *parametric*. The tests described in this chapter assume that gathered samples are randomly drawn from normal populations. They are strictly valid for data that are measured at least at the interval scale (see Sect. 1.2.2). Parametric tests about means usually involve more assumptions than that above. On the other hand, *nonparametric* methods are less restrictive than are parametric techniques and would, for example, possess utility in the analysis of market research questionnaires where interval and ratio measurement are rare. The chi square statistic for independence illustrated in Sect. 3.4 is an example of a nonparametric statistical test. Further nonparametric methods are described in the next chapter.

This chapter presents three widely used parametric statistical techniques that are used to make inferences about the mean(s) of populations from which samples have been taken. After a brief review of hypothesis testing, the second section of this chapter presents *the paired t test*. This is applied when we have two samples and in many applications of this test, the same cases appear in both samples. For example, we may assess workers' performance efficiencies before and after training. We would thus obtain two samples of performance scores, one pre-and the other post-training. The same individuals appear in the two samples. The samples are said to be paired or matched. Another example would be an examination of consumers' attitudes towards a product before and after their seeing some promotional material. The 'before-after' situation is a typical application of the paired t test.

The third section describes *the two-sample t test*. The difference between the two-sample t test and the paired t test is that the former is used when we have two

© Springer International Publishing Switzerland 2016
A. Aljandali, *Quantitative Analysis and IBM® SPSS® Statistics*,
Statistics and Econometrics for Finance, DOI 10.1007/978-3-319-45528-0_5

samples that are 'independent'. For example, we may take samples of two competing brands of computer and record a measure of performance for each computer in the samples. On the basis of the sample evidence, the objective would be to infer if there is a difference in the two population performance levels. Unlike the paired t test, the samples do not have to be of equal size. If we have more than two samples, for example, if we were examining performance levels of five brands of computer, then we turn to a technique called *analysis of variance* (ANOVA). There are many forms of analysis of variance and just the one-way ANOVA is described in the fourth section of this chapter. The assumptions and rationale of all three tests are described throughout the chapter. Associated IBM SPSS Statistics output is presented and discussed.

5.1   A Review of Hypothesis Testing

On the basis of sample evidence, the purpose of hypothesis testing is to make inferences about population values. These values are called parameters. In this chapter, the focus is on one particular population parameter—the mean. However, one may also make inferences about other population parameters such as the variance, as we saw with the Levene statistic of Sect. 3.2. The procedure involves common stages for most hypothesis tests:

- We set up a hypothesis of 'no difference' which is called the *null hypothesis* (denoted by H_0). We also set up an *alternative hypothesis* (denoted by H_1), to which we turn if we find that the null hypothesis is unlikely to be true.
- Under the assumption that the null hypothesis is true, we compute a *test statistic*. The test statistic has a numerical value and is computed using the sample information.
- We compute the probability of obtaining a test statistic at least as extreme as the one obtained, under the assumption that the null hypothesis is true. This probability is called the *significance*. In the case of hypothesis tests concerning the mean, the computation of this probability is often dependent upon the results of the Central Limit Theorem. (This theorem states that even when the study variable is not normally distributed, the sample mean will tend to be normally distributed. The larger the sample, the closer the sample mean is to being normally distributed.)
- If this computed probability is small (conventionally, less than 0.05 for a one-tailed test or less than 0.025 for a two-tailed test—see below), then the null hypothesis is rejected; otherwise, we fail to reject the null hypothesis.

Hypothesis tests may be *one or two-tailed*. If H_1 anticipates a 'greater than' or 'less than' result, the hypothesis test is always one-tailed, because the direction (+ or −) of the test statistic can be gauged. Alternative hypotheses of the 'do not equal' variety are two-tailed, as we cannot assess the sign of the test statistic.

5.2   The Paired t Test

SHOPPING.SAV contains retail information for a sample of 20 North American states. The following variables are contained in the file:

- STATE—a string variable containing the name of the state,
- YEAR2010—planned shopping centre square footage per capita in 2010 and
- YEAR2012—planned shopping centre square footage per capita in 2012.

The planned shopping centre has become the dominant component of the American retailing landscape. Accompanying the rapid expansion of these types of shopping centre has been a marked decline in the retail function of the central business districts of cities and towns of all sizes. This has resulted in a disadvantage for certain segments of the population, such as non-car owners unable to take advantage of the economies of scale offered by the planned shopping centres on the urban periphery.

The purpose of our investigation is to examine the rate of increase of planned shopping centres over time and to pinpoint inter-state variations. We are going to examine the growth in planned centres in just the 2-year period from 2010 to 2012. In particular, is the growth statistically significant over this relatively small time period?

5.2.1   Computation of the Test Statistic for the Paired t Test

In our example, we have two samples—YEAR2010 and YEAR2012—and the same states appear in both samples. The samples are thus paired. The paired t test uses the differences (d_i) between the paired sample figures. For illustration, the first three states in the data file and their paired differences are:

Shopping in North America

STATE	Alabama	Alaska	Arkansas
YEAR2010	20.95	19.23	21.23
YEAR2012	22.31	19.08	22.41
Differences (d_i)	−1.36	0.15	−1.18

The d_i for all 20 states are computed. It does not matter which way round the subtraction is performed to obtain the differences. Denoting the population mean difference by, the null and alternative hypotheses to be tested are:

H_0: The shopping centre square footages of all states in 2010 and 2012 are equal which implies that $\mu_d = 0$
versus

H_1: The population shopping centre square footages of 2012 exceed those of 2010 which implies that $\mu_d < 0$ if the subtraction is performed as YEAR2010 – YEAR2012.

Under the above null hypothesis, we would expect the sample mean of the differences, denoted by $\bar{d}$, to be close to zero. For our sample, $\bar{d} = -0.97$. We proceed to compute a test statistic using our numerical value of $\bar{d}$. The test statistic under H_0 is distributed as a t statistic with n – 1 degrees of freedom. This t statistic assumes that the d_i are drawn from a population of differences that are normally distributed. This assumption could be examined via the Shapiro–Wilks test.

5.2.2   The Paired t Test in IBM SPSS Statistics

To access the paired t test in IBM SPSS Statistics, from the Data Editor click:

Analyze
   Compare Means
      Paired-Samples T Test…

which generates the *Paired-Samples T Test dialogue box.* The variables YEAR2010 and YEAR2012 are selected (clicked) from the variable list and entered into the box headed 'Paired Variables' by the arrow button as per Fig. 5.1. Click the Options… button to produce the *Paired-Samples T Test: Options dialogue box* of Fig. 5.2. Here, the default is to produce a 95 % confidence interval for the difference between population means. Such an interval is constructed on the basis of the sample data, so that there is a probability of 0.95 that the difference between the population means is included in the interval. Click the Continue button to return to the *Paired-Samples T Test dialogue box* then click the OK button to derive the output of Fig. 5.3

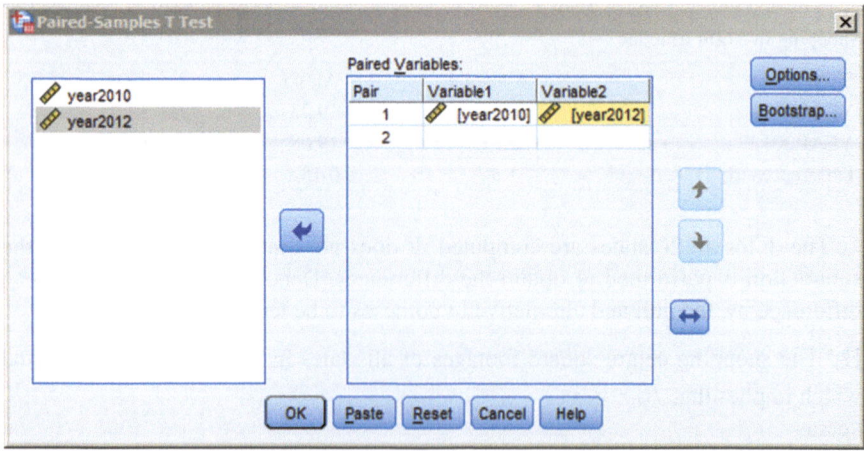

Fig. 5.1 The Paired-Samples T Test dialogue box

Fig. 5.2 The Paired-
Samples T Test: Options
dialogue box

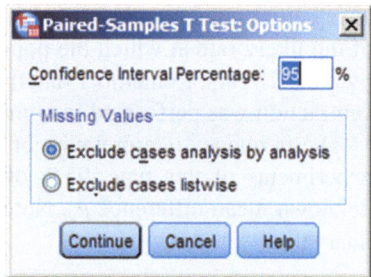

Paired Samples Statistics

		Mean	N	Std. Deviation	Std. Error Mean
Pair 1	year2010	23.3180	20	3.93227	.87928
	year2012	24.2880	20	3.80746	.85137

Paired Samples Correlations

		N	Correlation	Sig.
Pair 1	year2010 & year2012	20	.984	.000

Paired Samples Test

		Paired Differences							
					95% Confidence Interval of the Difference				
		Mean	Std. Deviation	Std. Error Mean	Lower	Upper	t	df	Sig. (2-tailed)
Pair 1	year2010 - year2012	-.97000	.71040	.15885	-1.30248	-.63752	-6.106	19	.000

Fig. 5.3 Results of applying the paired T Test to shopping centre data

which indicates that the mean of the d_i is $\overline{d} = -0.97$. The best estimate of the population standard deviation of the d_i appears under the heading 'Std. Deviation' and here has numerical value of 0.7104.

The test statistic has a value of -6.106 and is distributed as the t statistic with $n - 1 = 19$ degrees of freedom. In the present example, IBM SPSS Statistics has computed the differences by YEAR2010 $-$ YEAR2012. If it had been done the other way round, the test statistic would have been $+6.11$ because $\overline{d}$ would have been $+0.97$. Due to the symmetry of the t distribution, this would not affect the interpretation of the results. Our test is one-tailed test due to the nature of H_1. To obtain the significance of a test statistic for a one-tailed test, the user should divide the two-tailed probability by 2. In our example, the significance of the test statistic is zero to three decimal places. This means that the probability of obtaining a test statistic with numerical value of -6.11 or more is 0.000 to three decimal places, under our null hypothesis.

We, therefore, reject H_0 in favour of H_1 and conclude that population square footage for 2012 is in excess of that for 2010. A strictly more accurate interpretation is that the population mean of the differences, d_i, is in excess of zero. Even over such a short time span, the growth in planned shopping centre square footage is significant, reinforcing the dominance of this type of retail provision.

The *confidence interval* reported in Fig. 5.3 provides some information as to the extent of the difference in population mean square footages. The interval $-1.3025 < \mu_d < -0.6375$ has a probability of 0.95 of including the correct difference between the population mean difference. This interval gives some indication of the likely rate at which the population square footage per capita has increased over the 2 years. Remember that the minus signs are due to the way round that the subtraction was performed in computing the d_i.

The frequency interpretation of this type of inequality is that in the long run of experiments of this type, 95 % of them will yield an interval that contains the unknown mean difference μ_d. Note that $\mu_d = 0$ is not contained in this confidence interval.

5.3   The Two Sample t Test

STATE INCOME.SAV will be used to illustrate the two sample t test and the analysis of variance of the next section of this chapter. The file contains information about the per capita income in 2007 of a sample of 26 states in North America. The states in the sample fall into four geographical regions; northeast, north central, south and west.

The purpose is to pinpoint any significant differences in the population mean per capita income levels between these four regions. It should be appreciated that the analysis is, therefore, at the macro-level and any micro-level variations in this variable may be masked. The file contains the following three variables that have already been labelled, as have the codes that define the four geographical regions:

- STATE—a string variable containing the name of the states in the sample,
- INCOME—the dollar per capita income levels and
- REGION—a nominal variable that defines the geographical regions. The codes used were 1 for 'northeast', 2 for 'north central', 3 for 'south' and 4 for 'west'.

The two sample t test will be used to test for a significant difference in the mean per capita income levels between the southern and western regions.

5.3.1   Computation of the Test Statistic for the Two Sample t Test

If we denote the sample mean per capita income levels in southern states as $\overline{X}$ and that in the western states by $\overline{Y}$, then the two sample t test will be used to examine the null and alternative hypotheses:

H_0: the population income levels are the same in southern and western states $\mu_x = \mu_y$ or $\mu_x - \mu_y = 0$.
versus
H_1: the population income levels in these two regions are unequal, $\mu_x = \mu_y$ or $\mu_x - \mu_y = 0$.

Computation of the test statistic for the two sample t test depends on the following assumptions:

(1) The samples were selected independently and randomly,
(2) The samples have both been gathered from normal populations and
(3) The samples have been taken from populations with equal variance. This property is called *homoscedasticity*.

The consequence of markedly violating these assumptions is that exact levels of significance cannot be computed. As has previously been discussed, IBM SPSS Statistics has made every attempt to provide facilities for examining such assumptions as the above. The numerical input to the test statistic associated with this test is the difference between the two sample means i.e. $\bar{X} - \bar{Y}$. Ideally, under the null hypothesis, we would expect this difference to be zero. As required by the third assumption above, we do not know the supposed equal population variance. It has to be estimated. The best estimator of this common variance is called the *pooled estimator* and is reported in the IBM SPSS Statistics output. The test statistic for the two sample t test is again distributed as a t statistic, but this time with $n_X + n_Y - 2$ degrees of freedom and where n_X and n_Y are the respective numbers of readings in the two samples.

By default, IBM SPSS Statistics presents the Levene test discussed in Sect. 2.2.1 as part of the output to the two sample t test. It will be recalled that the Levene test examines the hypothesis that the samples are drawn from populations with equal variance and so assesses whether our third assumption above is reasonable. If the Levene test evidences that the homoscedasticity assumption is violated, then we turn to what is called the *Behrens–Fisher* version of the two sample t test, which is designed for the situation of unequal population variances.

5.3.2 The Two Sample t Test in IBM SPSS Statistics

To access the two sample t test in IBM SPSS Statistics, from the Data Editor click:

Analyze
    Compare Means
        Independent-Samples T Test…

which gives rise to the *Independent Samples T Test dialogue box* of Fig. 5.4. In the box headed 'Test Variable' enter the variable INCOME in the usual manner. The variable REGION is our 'Grouping Variable', so enter this variable into the box. The two question marks indicate that IBM SPSS Statistics needs to know the codes that define our regions. Reference to the start of Sect. 5.3 reminds us that a code of 3 represents the 'south' and 4 represents the 'west', which are the two regions to be compared. Note that the Define Groups… button has become active now that REGION has been entered. Click this button to obtain the Define Groups dialogue box of Fig. 5.5. Simply enter the code of 3 in the box labelled Group 1. Click inside the box labelled Group 2 and type in the code of 4. Click the Continue button to return to the dialogue box of Fig. 5.4. Now click the Options… button to obtain dialogue box of Fig. 5.6. The default of a 95 % confidence interval applies to the difference between population means.

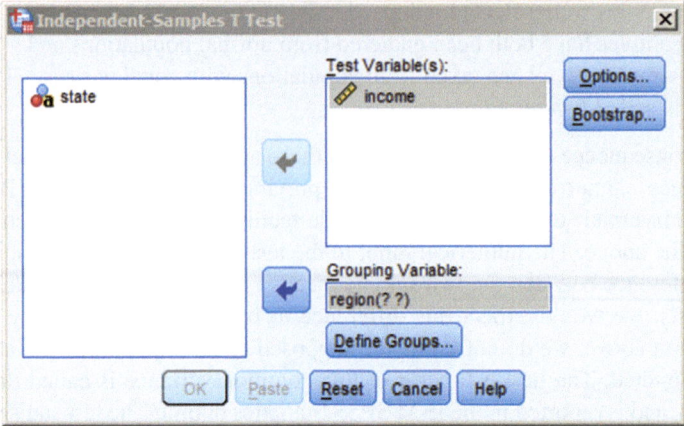

Fig. 5.4 The Independent Samples T Test dialogue box

Fig. 5.5 The Define
Groups dialogue box

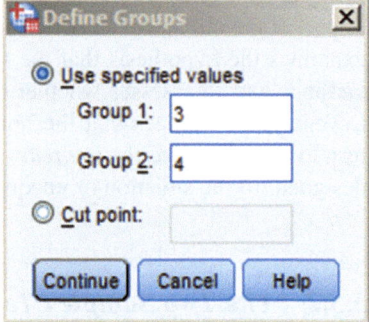

Fig. 5.6 The Independent
Samples T Test: Options
dialogue box

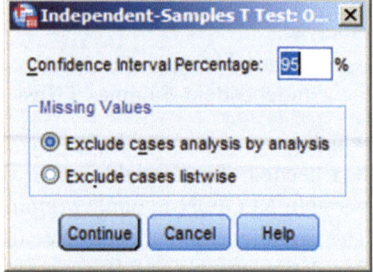

Click the Continue button to return to Fig. 5.4 and click the OK button to run this statistical test. The results of applying the two sample t test to our per capita income data are presented in the IBM SPSS Statistics Viewer of Fig. 5.7.

The difference between the sample means is $W = \bar{X} - \bar{Y} = -2732.444$. This difference may seem numerically large, so the null hypothesis could be in doubt. However,

Group Statistics

	GEOGRAPHICAL REGION	N	Mean	Std. Deviation	Std. Error Mean
PER CAPITA INCOME, 2007	SOUTH	9	51535.56	6628.522	2209.507
	WEST	5	54268.00	13515.364	6044.255

Independent Samples Test

		Levene's Test for Equality of Variances		t-test for Equality of Means					95% Confidence Interval of the Difference	
		F	Sig.	t	df	Sig. (2-tailed)	Mean Difference	Std. Error Difference	Lower	Upper
PER CAPITA INCOME, 2007	Equal variances assumed	1.557	.236	-.516	12	.615	-2732.444	5296.788	-14273.155	8808.266
	Equal variances not assumed			-.425	5.095	.688	-2732.444	6435.444	-19182.951	13718.062

Fig. 5.7 Output generated by the Independent samples T test

at this stage this is a fallacious conclusion. Significance or otherwise also depends on the relative spread inherent in the data. The significance associated with the Levene test statistic (denoted by F) is $p=0.236>0.05$. This is not significant, so we fail to reject the hypothesis that the population variances are equal. The homoscedasticity assumption is tenable, so we read the output for 'equal variances assumed' in Fig. 5.7. If the homoscedasticity assumption is not tenable, then we would turn to the t test results for 'equal variances not assumed'. (Note that the Behrens–Fisher version of this test can have non-integer degrees of freedom!)

The test statistic is has numerical value -0.516 and is distributed as a t statistic with 12 $(9+5-2)$ degrees of freedom. From Fig. 5.7, the significance of this test statistic is $p=0.615>0.025$. We, therefore, fail to reject the null hypothesis and conclude that the population mean incomes per capita in the southern and western regions are equal. Examination of the confidence interval indicates that there is a probability of 0.95 that the interval $-14273.15 < \mu_X - \mu_Y < 8808.27$ includes the correct difference between population means.

As expected from the result of the above hypothesis test, a value of zero is included in this interval. However, the width of this interval does not make it particularly useful. We really need larger sample sizes. The Behrens–Fisher version of the test is irrelevant here, due to the equality of variance assumption being met. However, for completeness, let us run through the IBM SPSS Statistics output to see where various quantities have come from. This test statistic has numerical value -0.425 with a significance level of 0.688 and the conclusion would have been as before, should the Behrens–Fisher test been relevant.

5.4   The One-Way Analysis of Variance

The one-way analysis of variance (*one-way* or *one-factor ANOVA*) technique is used to detect differences between the population means of more than two groups. For example, we may wish to test differences in crop yields over eight different brands of fertiliser. One may think that this could be accomplished by performing two sample t tests for every pair of fertiliser. This would necessitate 28 separate t tests. However, it is conventional in hypothesis tests to reject H_0 when the significance is less than 0.05. This is because there is less than a 0.05 probability of obtaining the test statistic or one more extreme, under the particular H_0 in question. Thus, in performing 28 t tests, there would be a reasonable chance of rejecting H_0 at least once, when in fact H_0 should not have been rejected. This is why the ANOVA method should be used when there are more than two groups. It should be noted that a *two-way* or *two-factor ANOVA* would involve two classifying variables. For example, we may wish to examine crop yields over eight brands of fertiliser and over three types of soil. One could simultaneously look for significant differences in yield over (a) brands and (b) soil types.

The data in the file CAPITA INCOME.SAV will be used to illustrate the one-way ANOVA. We have four geographical regions. The one-way ANOVA is used to assess if the population mean per capita income levels are equal across these four regions.

5.4.1   Computation of the Test Statistic for the One-Way ANOVA

Once more, being a parametric technique, there are assumptions upon which computation of the ANOVA test statistic depends. The assumptions underlying the parametric ANOVA are:

(1) Each group is an independently selected random sample,
(2) Each group contains sample data drawn from a normal population and
(3) The data in each group have been drawn from populations that have equal variance.

Returning to our data file CAPITA INCOME.SAV, we shall use a one-way ANOVA to test the following null and alternative hypotheses:

H_0: the population mean per capita income levels are the same for the four regions $\mu_1 = \mu_2 = \mu_3 = \mu_4$.
Versus
H_1: one or more of these means differ

The test statistic under the null hypothesis is distributed as F. It should be noted that the F statistic is vulnerable to departures from normality and this assumption should be checked. If the F statistic is significant and H_0 is rejected, we know that the population means are not all equal. To pinpoint which pairs of means are significantly different, IBM SPSS Statistics presents several *multiple comparison procedures (MCP)*. Many such procedures are available, but they all tend to adjust for the number of comparisons between groups that have to be made. Essentially, if many such comparisons are made, the larger must be the difference between pairs of means for the multiple comparisons procedure to find it significant. Research on the various MCP methods available shows that many are similar in respect of their power.

5.4.2   The One-Way ANOVA in IBM SPSS Statistics

To access the IBM SPSS Statistics one-way ANOVA routine, from the Data Editor click:

Analyze
    Compare Means
        One-way ANOVA…

which produces the *One-Way ANOVA dialogue box* of Fig. 5.8. Click the variable INCOME and enter into the box titled 'Dependent List'. The 'Factor' in this case is the geographical region, so click REGION into the relevant box.

Clicking the Options… button in the dialogue box of Fig. 5.8 gives rise to the *One-Way ANOVA: Options dialogue box* of Fig. 5.9. It is wise to select (click) the

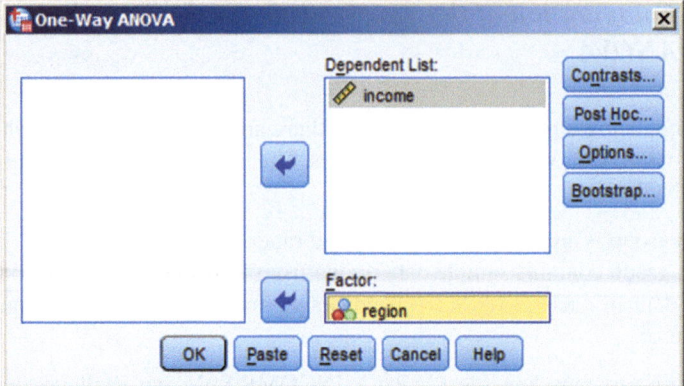

Fig. 5.8 The One-Way ANOVA dialogue box

Fig. 5.9 The One-Way
ANOVA: Options dialogue
box

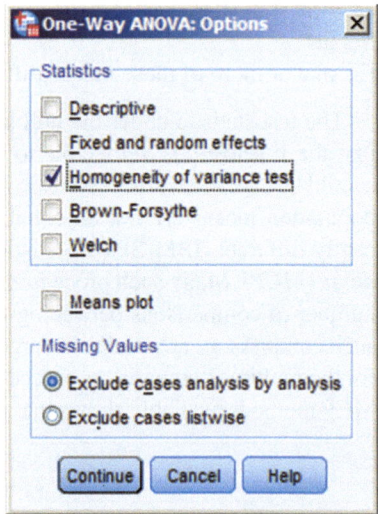

option 'Homogeneity-of-variance' which generates the Levene test. Given that homoscedasticity is an assumption of the one-way ANOVA, this should be assessed. Also the Descriptive option may provide useful information. For example, examination of the group means may indicate between which groups any significant differences lie. Click the Continue button in the above dialogue box to return to the dialogue box of Fig. 5.8.

Lastly, click the button labelled Post Hoc... in Fig. 5.8 which generates the *One-Way ANOVA: Post Hoc Multiple Comparisons dialogue box* of Fig. 5.10. This dialogue box permits multiple comparisons, which of course are only of interest if we reject the null hypothesis. In this dialogue box, *the Scheffe test* has been selected, because it requires larger differences between sample means for significance than do the other optional tests.

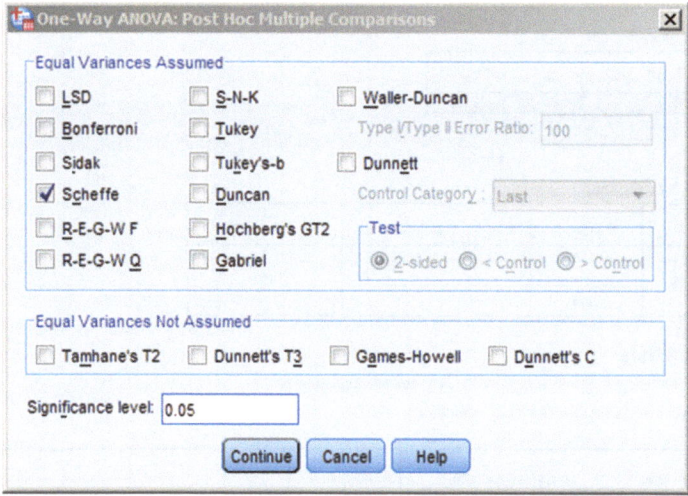

Fig. 5.10   The One-Way ANOVA: Post Hoc Multiple Comparisons dialogue box

Click the Continue button to return to the dialogue box of Fig. 5.7 and click the OK button to run the one-way ANOVA, for which the major part of the results are presented in the *IBM SPSS Statistics Viewer* of Fig. 5.10.

5.4.3   Discussion of the Results of the One-Way ANOVA

Figure 5.11 presents some of the descriptive statistics produced by the one-way ANOVA, if this selection is made from the *One-Way ANOVA: Options dialogue box* of Fig. 5.9. Examination of the four group means suggests not a great deal of difference between the four regions. However, one notices that the four sample standard deviations do seem different and this could cast doubt on the homoscedasticity assumption of the technique. The north central region shows much less spread of incomes than do the other three regions. However, the Levene test statistic is not significant ($p=0.097>0.05$), so we fail to reject the hypothesis of equal population variances.

The F statistic has numerical value of 0.921, with associated significance $0.447>0.05$. We thus fail to reject the null hypothesis that the population mean per capita incomes are equal across the four geographical regions. No groups are significantly different at the 0.05 level.

Having a non-significant test statistic, it is obvious that the Scheffe multiple comparisons procedure will not detect any significant differences in income levels between pairs of regions. This fact is in evidence in the final part of Fig. 5.11. For

Test of Homogeneity of Variances

PER CAPITA INCOME, 2007

Levene Statistic	df1	df2	Sig.
2.379	3	22	.097

ANOVA

PER CAPITA INCOME, 2007

	Sum of Squares	df	Mean Square	F	Sig.
Between Groups	175406269.5	3	58468756.51	.921	.447
Within Groups	1396494565	22	63477025.69		
Total	1571900835	25			

Post Hoc Tests

Multiple Comparisons

Dependent Variable: PER CAPITA INCOME, 2007

Scheffe

(I) GEOGRAPHICAL REGION	(J) GEOGRAPHICAL REGION	Mean Difference (I-J)	Std. Error	Sig.	95% Confidence Interval	
					Lower Bound	Upper Bound
NORTHEAST	NORTH CENTRAL	-3548.857	4665.142	.900	-17658.41	10560.69
	SOUTH	3118.444	4443.918	.919	-10322.02	16558.91
	WEST	386.000	5038.929	1.000	-14854.05	15626.05
NORTH CENTRAL	NORTHEAST	3548.857	4665.142	.900	-10560.69	17658.41
	SOUTH	6667.302	4015.115	.448	-5476.26	18810.87
	WEST	3934.857	4665.142	.869	-10174.69	18044.41
SOUTH	NORTHEAST	-3118.444	4443.918	.919	-16558.91	10322.02
	NORTH CENTRAL	-6667.302	4015.115	.448	-18810.87	5476.26
	WEST	-2732.444	4443.918	.944	-16172.91	10708.02
WEST	NORTHEAST	-386.000	5038.929	1.000	-15626.05	14854.05
	NORTH CENTRAL	-3934.857	4665.142	.869	-18044.41	10174.69
	SOUTH	2732.444	4443.918	.944	-10708.02	16172.91

Homogeneous Subsets

PER CAPITA INCOME, 2007

Scheffe[a,b]

GEOGRAPHICAL REGION	N	Subset for alpha = 0.05
		1
SOUTH	9	51535.56
WEST	5	54268.00
NORTHEAST	5	54654.00
NORTH CENTRAL	7	58202.86
Sig.		.554

Means for groups in homogeneous subsets are displayed.

a. Uses Harmonic Mean Sample Size = 6.117.

b. The group sizes are unequal. The harmonic mean of the group sizes is used. Type I error levels are not guaranteed.

Fig. 5.11 Output from the one-way analysis of variance procedure

example, the difference in sample mean income levels between the Northeast and North Central regions is -$3548.86 and is non-significant (p=0.900); the difference in sample means between the Northeast and Southern regions is $3118.44, again non-significant (p=0.919).

Chapter 6
Nonparametric Hypothesis Tests

The parametric methods of the previous chapter required data measured at the interval or ratio levels and which is normally distributed. Business data are not always at these levels of measurement. Market research regularly produces data at the *nominal* (e.g. "agree" versus "disagree" with a proposition about a product) and *ordinal* (e.g. ranked preferences) levels. The study of consumer preference is a field in which data at nominal and ordinal levels are particularly evident. In such instances the analyst has recourse to nonparametric statistical methods. Serious doubts about the normality assumption even when the data are at the interval or ratio levels is another situation in which nonparametric methods may be preferred over parametric ones. Many authors refer to nonparametric methods as *distribution free*, in that they make relatively few assumptions about the nature of the population distribution.

In Statistics, the *power* of a hypothesis test is defined as its ability to reject the null hypothesis when indeed it should be rejected. Obviously, statistical tests should have high power. Nonparametric hypothesis tests often possess almost as much power as do parametric tests, when the normality and other assumptions are satisfied. The former are often more powerful in detecting population differences when the required assumptions are not satisfied.

The nonparametric tests presented in this chapter parallel the parametric tests presented in the previous chapter. Of the numerous nonparametric tests available, this chapter has been selective in presenting methods that have ready business application and high power. The first section presents one of the oldest and most widely used statistical tests—*the sign test*. This is applicable when we have paired or matched samples. The methodology underlying the sign test depends on whether we have relatively small or large samples.

The second section considers two independent samples. There are many nonparametric methods for examining the differences between two independent groups. Such independent samples may arise in two situations. Firstly, the two samples may be drawn at random from two populations. Secondly, the samples may be generated by the random assignment of two treatments to the members of some sample whose

© Springer International Publishing Switzerland 2016
A. Aljandali, *Quantitative Analysis and IBM® SPSS® Statistics,*
Statistics and Econometrics for Finance, DOI 10.1007/978-3-319-45528-0_6

origins are arbitrary. For example, we may randomly assign individuals in a consumer panel to one of two different forms of product advertisement (treatments) and seek their evaluation of the medium to which they have been exposed. When we wish to test for differences in central tendency and have at least ordinal measurement, the *Mann–Whitney* test has good power. (There is a nonparametric test called *the permutation test* that could be used to examine differences in central tendency and which has more power, but it is only applicable to small samples.)

The third section involves the situation in which we have more than two independent samples. We thus turn to nonparametric analysis of variance. Of the available nonparametric ANOVA methods, *the Kruskal–Wallis test* is the most efficient in that it uses more of the information available in the sample readings. This test also has an associated multiple comparisons procedure.

6.1   The Sign Test

EXPENDITURES.SAV permits examination of patterns of UK household expenditure for nine commodities at two different dates—2014 and 2015. Average annual household expenditures for each commodity are converted to percentages of overall annual household expenditure. For example, in 2014, an average of 13.47% of household expenditure went on food in the UK; in 2015 the comparable figure was 12.80%. The data file contains the following three variables which have already been labelled for you:

- COMMOD—a string variable defining each of the nine commodities,
- Stat2014—average expenditure/commodity as a percentage of total annual household expenditure for 2014 and
- Stat2015—average expenditure/commodity as a percentage of total annual household expenditure for 2015.
- The sign test is to be used to examine if the population distributions of household expenditures are the same at 2014 and 2015 levels.

6.1.1   Computation of the Test Statistic for the Sign Test

The sign test makes no assumptions about the population distributions of expenditures. The null and alternative hypotheses are:

H_0: the population distributions of expenditures are the same, versus
H_1: the two population distributions differ.

Note that the alternative hypothesis does not specify the nature of the difference. The populations may differ in central tendency and/or spread, for example. The sign test derives its name from the fact that the test statistic is based on the direction of differences between the two variables at hand. For example, four of the commodities in EXPENDITURES.SAV are reproduced below:

Household expenditures

Sign of COMMOD	stat2014	stat2015	Difference
Housing	44.9	45.0	–
Transport	7.9	7.8	+
Food	13.5	12.8	+
Household good and services	5.3	5.3	O

If we consider the difference stat2014 – stat2015, then this difference is negative for housing, but positive for food and transport. In the latter case, this simply means that average UK household expenditures on food and transport in 2014 were in excess of the comparable expenditures for 2015. The household goods and services proportions are the same for both years. We thus have *a tie* as exemplified by the 'O' in the above table. Note that the subtraction could be performed the other way round; it would not influence the significance of the test statistic.

Under the null hypothesis, we would expect the same number of positive and negative differences in the sample. Conversely, if too few differences of one sign occur, then the null hypothesis would be rejected. (Technically what we are saying is that if the null hypothesis is tenable then the median of the differences between stat2014 and stat2015 is zero.)

Under H_0, the probability of obtaining a positive (or negative) difference is 0.5. If we denote the number of plusses (or minuses) obtained in a sample by X, then we can compute the probability of gaining X or more plusses (or minuses) out of a total of n plusses (or minuses) by the binomial distribution and where n is the sample size. For example, suppose in our sample of $n = 11$ commodities, we obtained $X = 9$ positive differences, then under H_0 with the probability of gaining a plus as 0.5, the binomial distribution gives:

$$P(X \geq 9) = P(X=9) + P(X=10) + P(X=11)$$
$$= 55(.5)^9 (.5)^2 + 11(.5)^{10} (.5)^1 + (.5)^{11}$$
$$= 67(.5)^{11} = 0.0327$$

For a two tailed test, the null hypothesis is rejected if this probability is less than 0.025, so in the above hypothetical example, we would have failed to reject H_0. Note that if there are ties, then the associated observations are removed from further analysis and n is reduced accordingly. If in the above example, we had let X represent the number of minuses gathered, then X would have equalled 2.

If we compute the probability of obtaining a value of X of 2 or less, that probability would again be 0.0327. This is because the binomial distribution is symmetrical when the probability of interest equals 0.5. Hence, it matters not whether X represents the number of plusses or minuses obtained.

Under certain conditions, the binomial distribution is adequately approximated by the normal distribution. A rule of thumb is that this approximation is adequate of one of the criteria below is met. Representing the probability of interest as p (and in our example, $p = 0.5$):

- If $p < 0.5$, np should be greater than 5 or
- If $p \geq 0.5$, $n(1-p)$ should be greater than 5.

In such instances, IBM SPSS Statistics will report a Z score, where Z is a *standard normal deviate* (mean of zero, variance of unity). Again symmetry applies. If we let X denote the number of plusses obtained, we might obtain an associated Z score of say 1.87 in a particular experiment; if we had focussed on letting X equal the number of minuses obtained in that experiment, then the Z score would be −1.87 in this fictitious example. The level of significance remains the same.

6.1.2   The Sign Test in IBM SPSS Statistics

Nonparametric statistical tests are accessed by clicking from the Data Editor:

Analyze
   Nonparametric Tests
      Legacy Dialogs…

which gives rise to a series of choices. In that our expenditure data involves two related or paired samples, we select (click):

2 Related Samples…

which generates the *Two Related Samples Tests dialogue box* of Fig. 6.1. It will be noted that along with the option of choosing the sign test are the choices labelled Wilcoxon and McNemar for two related samples. (The Wilcoxon test for two related samples is often called the Wilcoxon Signed Ranks test in the literature.)

In the box labelled 'Test Type' choose (click) the Sign test and deselect (click) the default of the Wilcoxon test. The variables under examination are stat2014 and

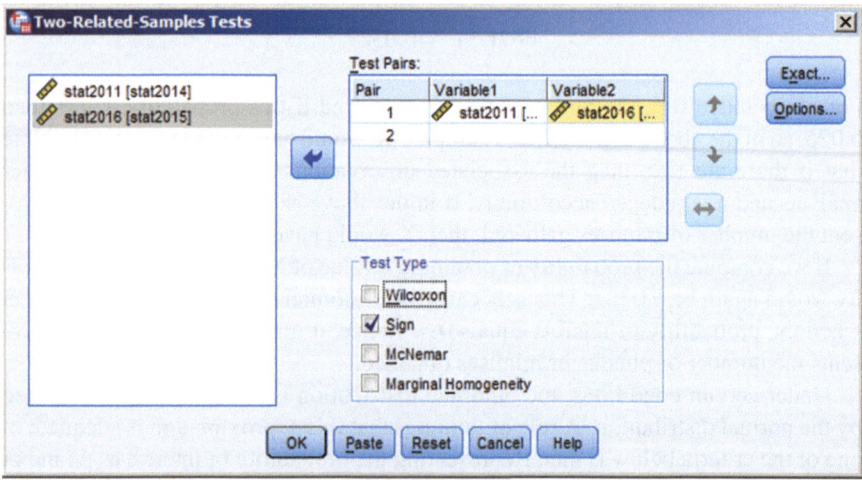

Fig. 6.1 The Two Related Samples Tests dialogue box

stat2015, so these are chosen from the variable list and entered into the box titled 'Test Pairs' via the arrow. At the bottom of the *Two Related Samples Tests* dialogue box is the Options… button. Upon clicking this, the *Two Related Samples: Options dialogue box* of Fig. 6.2 is obtained. This is self-explanatory, in that various Descriptive statistics and Quartiles are available. Click the Continue button to return to the previous dialogue box and now click the OK button to run the sign test. The results of the sign test are presented in the IBM SPSS Statistics Viewer of Fig. 6.3.

Fig. 6.2 The Two Related Samples: Options dialogue box

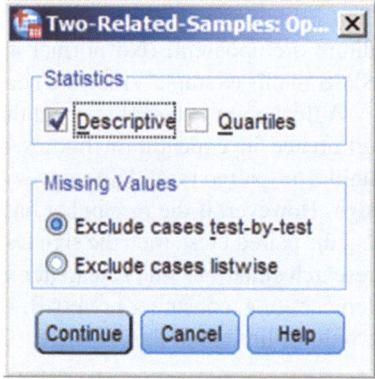

Fig. 6.3 Output from the sign test

Descriptive Statistics

	N	Mean	Std. Deviation	Minimum	Maximum
stat2011	9	9.9889	13.63758	1.20	44.90
stat2016	9	9.9444	13.63691	1.20	45.00

Sign Test

Frequencies

		N
stat2016 - stat2011	Negative Differences[a]	2
	Positive Differences[b]	4
	Ties[c]	3
	Total	9

a. stat2016 < stat2011

b. stat2016 > stat2011

c. stat2016 = stat2011

Test Statistics[a]

	stat2016 - stat2011
Exact Sig. (2-tailed)	.687[b]

a. Sign Test

b. Binomial distribution used.

The output below reminds us that three commodities (health goods & medical services, household goods & services and communications) have been eliminated from the analysis due to tied values for 2014 and 2015. This leaves six commodities and two negative differences were obtained and, therefore four positive differences. From the information under the 'Frequencies' heading, the method of subtraction used by IBM SPSS Statistics here was stat2015 − stat2014, but remember that the method of subtraction is irrelevant. The binomial probability of obtaining 5 or more plusses under H_0 (or, if you prefer, 5 or less minuses) is 1.000 to three decimal places. We, therefore, fail to reject the null hypothesis and conclude that the population distributions of UK weekly expenditures on these commodities are the same. This is reinforced at the top of Fig. 6.3, where the similarities in patterns of expenditure are apparent. (No normal approximation to the binomial is used by IBM SPSS in this example, since the necessary criterion is not met.)

A final point to be noted about the sign test is that it only uses the direction of the difference in expenditure figures and ignores the magnitude of this difference, unlike the paired t test. Differences of 0.1 or 51 % in expenditures generate the same sign. However, if the researcher has serious doubts about the assumptions underlying the paired t test, then the sign test is very likely to possess more power. In market research situations, the researcher may only have available the direction of the differences (e.g. consumer prefers Brand X to Brand Y), so alternatives to the sign test would be limited in this instance.

6.2   The Mann–Whitney Test

Before any discussion of this test for two independent samples, it must be noted that it is often referred to as the Mann–Whitney–Wilcoxon test in the literature. (Do not confuse this with the Wilcoxon Signed Ranks test alluded to in the previous section and which is used for two related samples.) This different name is because these three statisticians independently proposed nonparametric tests which are essentially the same. IBM SPSS Statistics calls the test described in Sect. 6.2.1 below the *Mann–Whitney test*; other authors call it the Wilcoxon test.

The null hypothesis of the Mann–Whitney test is that the samples have been drawn from populations with the same distribution. The two sample t test of the previous chapter focused on differences between population means, rather than differences in dispersion or differences in distributional form. As the null hypothesis suggests, the Mann–Whitney procedure essentially tests for all of these possible differences simultaneously. However, if the research focus is to test whether two samples represent populations that differ in central tendency, then the Mann–Whitney test is amongst the most sensitive to such differences.

The Mann–Whitney test is illustrated by an example taken from the field of consumer behaviour. In studies of consumer behaviour it has been noted that in some purchasing situations, shoppers compare several alternatives before making a purchase. In particular, this applies to purchases that the consumer in some way deems as "important". The number of alternatives considered prior to purchase is called the buyer's *evoked set* in Marketing. These alternatives are defined according to the study objective, but in the retailing field could refer to different brands of a product, competing shops selling a product or even to different shopping centres/retail parks offering a particular product. Retailers want to be part of the consumer comparison process and retail-oriented studies have established several factors as having an influence on the size of the consumer's evoked set in varying contexts. One such (psychological) factor is the degree of risk that shoppers attach to the purchase at hand. This has been referred to as *perceived risk*. For example, consumers perceiving high levels of risk attached to a particular purchase may be hypothesized to consider more alternatives in order to reduce the risk factor.

The data file RISK.SAV contains information about 150 consumers who were asked which shops they typically compare before purchasing shoes in a business district. By means of a well-tested measuring device, their levels of perceived risk attached to this purchase were classified as 'low', 'medium' or 'high'. Two of the variables in this IBM SPSS Statistics data file are:

- ESET—the size of each consumer's evoked set and
- PERRISK—levels of perceived risk, coded as 1 for 'very low' up to 5 for 'very high'.

The Mann–Whitney test will be applied to see if the evoked set sizes differ for consumers perceiving 'very low' and 'very high' levels of risk.

6.2.1   Computation of the Mann–Whitney Test Statistic

As stated above, the Mann–Whitney test adopts the null hypothesis that the two samples have been drawn from populations with the same distribution. Suppose that we have two samples of sizes n_X and n_Y. Let the total number of readings in both samples be $n = n_X + n_Y$. In computing the test statistic for the Mann–Whitney test, the first stage is to combine the scores of both samples and rank the observations from smallest (rank of 1) to highest (rank of n). Taking each score in one sample, we count the number of scores in the second sample that have higher ranks. The total of this count is denoted by U. For example:

Sample 1: 12 13 17
Sample 2: 14 18 19 20

would suggest that the null hypothesis is not tenable, in that the readings in the first sample are generally lower than those of the second sample. Combining the two samples and ranking from lowest to highest:

Combined data: <u>12</u> <u>13</u> 14 <u>17</u> 18 19 20
Rank: 1 2 3 4 5 6 7

For clarity, the readings from the first sample are underlined. Take each reading in Sample 1 and count the number of readings in Sample 2 that have higher ranks:

$$U = 4+4+3 = 11.$$

Conversely, if we had focused on Sample 2, take each reading in Sample 2 and count the number of readings in Sample 1 that have higher ranks:

$$U^* = 1+0+0+0 = 1.$$

A relatively large score for U (or a relatively low score for U*) would suggest that the readings have not been drawn from populations with the same distribution. If the null hypothesis is tenable, we would expect that for each of the n_X readings in Sample 1, one half of the n_Y readings in Sample 2 would have larger ranks. Thus, under H_0, the expected or mean value of U is:

$$\text{Mean (U)} = (n_X * n_Y)/2 \qquad (6.1)$$

In the above numerical example, mean (U) = 6. Note that U and U* are symmetrical about this mean (each five units distant from it). Either U or U* may be used as the test statistic and statistical tables are available for small sample sizes. IBM SPSS Statistics reports the smaller of U and U*. Also, the smaller of U and U* is typically tabulated in statistical tables. Whichever of U or U* is employed, a significance of less than 0.05 leads to rejection of H_0. For larger sample sizes, computation of U can be tedious. For larger samples, with both n_X and n_Y in excess of 10, an approximately standard normally distributed test statistic is derived for assessment of H_0.

The U and U* statistics are the Mann–Whitney versions of this test. The Wilcoxon version uses the sum of the ranks allocated to each sample. In our numerical example, $R_X = 7$ and $R_Y = 21$ are these respective sums of ranks. (It may be recalled that the sum of the first n integers—here ranks—is $n(n+1)/2$, so $R_X + R_Y$ must equal $(7 \times 8)/2 = 28$, even if there are tied ranks.) Under H_0, we would expect the average of the ranks for each sample to be about equal. Wilcoxon used the notation W instead of R_x for the sum of the ranks, and IBM SPSS Statistics preserves this notation.

6.2.2   The Mann–Whitney Test in IBM SPSS Statistics

Returning to the data file RISK.SAV, the null and alternative hypotheses to be tested are:

H_0: the evoked set sizes for consumers perceiving low (code of '1') and high (code of '5') levels of risk are drawn from populations that possess the same distribution, versus
H_1: The two samples are not drawn from such populations.

It would be reasonable to conclude in this context that rejection of the null hypothesis infers a difference in central tendency in the population evoked set sizes for these two groups.

From the Data Editor, click:

Analyze
    Nonparametric Tests
        Legacy Dialogs
            2 Independent Samples…

which produces the *Two Independent Samples dialogue box* of Fig. 6.4. The Mann–Whitney U test is the default amongst the four test options available. The variable ESET is entered into the 'Test Variable List' and PERRISK is the 'Grouping Variable'. Upon performing the latter task, the Define Groups… button becomes active, because we need to enter the codes that define our two groups. Click this button to obtain the *Two Independent Samples: Define Groups dialogue box* of Fig. 6.5. Our two codes of 1 and 5 define consumers perceiving very low and very high levels of risk, so these codes are typed into the appropriate boxes. Click the Continue

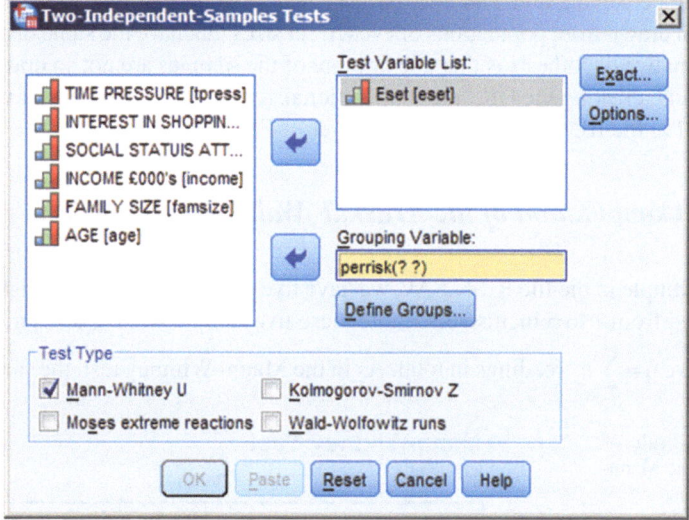

Fig. 6.4 The Two-Independent-Samples Tests dialogue box

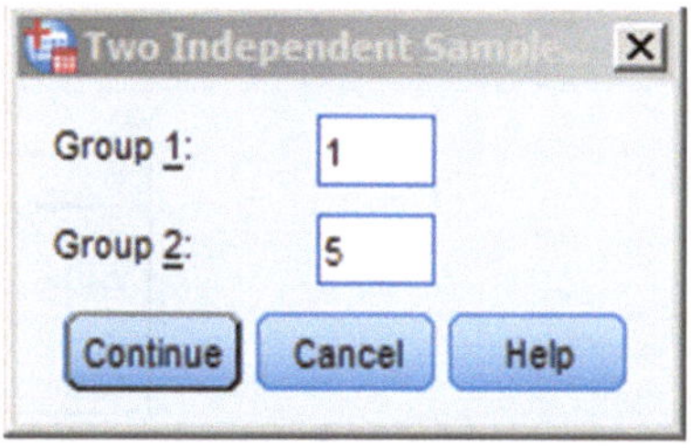

Fig. 6.5 The Two Independent Samples: define dialogue box

button to return to the dialogue box of Fig. 6.4. The Options… button gives rise to selection of descriptive statistics and quartiles, just as in the case of the sign test.

As anticipated, the evoked set sizes of consumers perceiving very low risk are smaller than those perceiving very high risk. This is indicated by the means of the ranks attached to these two groups. Shoppers perceiving high purchase risk consider more alternatives. Perception of risk is, therefore, a significant determinant of the extent of consumer search. The Wilcoxon W statistic is the smaller sum of ranks (i.e. for consumers perceiving very low risk).

6.3   The Kruskal–Wallis One-Way ANOVA

The Mann–Whitney test for two independent samples presented in the previous section, was extended to the problem of analysing K independent samples where K > 2, by Kruskal and Wallis. The test requires at least ordinal measurement and is generally more powerful than its nonparametric competitors. The data file RISK.SAV will again be used, but now introducing consumers who perceive medium levels of risk along with those perceiving high and low risk. The null hypothesis is that all three samples have been drawn from populations of evoked set sizes that have the same distribution. The alternative hypothesis is that one or more of the samples are not so drawn.

As usual, clicking the OK button operationalizes the test, the results of which are presented in the IBM SPSS Statistics Viewer of Fig. 6.6.

6.3.1   Computation of the Kruskal–Wallis Test Statistic

In our example in the file RISK.SAV, we have five groups or samples of risk perception coded from 1 to 5 inclusive. Denote these five sample sizes as sizes $n_1, n_2 \ldots n_5$ so we have $n = \sum_{i=1}^{s} n_i$ readings in total. As in the Mann–Whitney test, the samples are

Fig. 6.6 Results of applying the Mann–Whitney test

Mann-Whitney Test

Ranks

	PERCEIVED RISK	N	Mean Rank	Sum of Ranks
Eset	very low	48	24.50	1176.00
	very high	42	69.50	2919.00
	Total	90		

Test Statistics[a]

	Eset
Mann-Whitney U	.000
Wilcoxon W	1176.000
Z	-8.417
Asymp. Sig. (2-tailed)	.000

a. Grouping Variable: PERCEIVED RISK

combined into one sample and all of the observations ranked from lowest (rank of 1) to highest (rank of n) regardless of sample membership.

The logic of the Kruskal–Wallis test is that if the observations from all K samples are from populations with the same distribution, then the average of the ranks allocated to each sample should be about the same. Computation of the test statistic (commonly denoted by H) of the Kruskal–Wallis test employs the average of the ranks allocated to each sample. The test examines the differences in average ranks to assess if they are so disparate which indicates that they are unlikely to have been drawn from populations with the same distribution. When there are more than five readings in each group, H is approximately distributed as a chi square statistic with K − 1 degrees of freedom.

6.3.2 The Kruskal–Wallis Test in IBM SPSS Statistics

This test is accessed by clicking from the Data Editor:

Analyze
    Nonparametric Tests
        Legacy Dialogs
            K Independent Samples…

which gives rise to the *Tests for Several Independent Samples dialogue box* of Fig. 6.7.

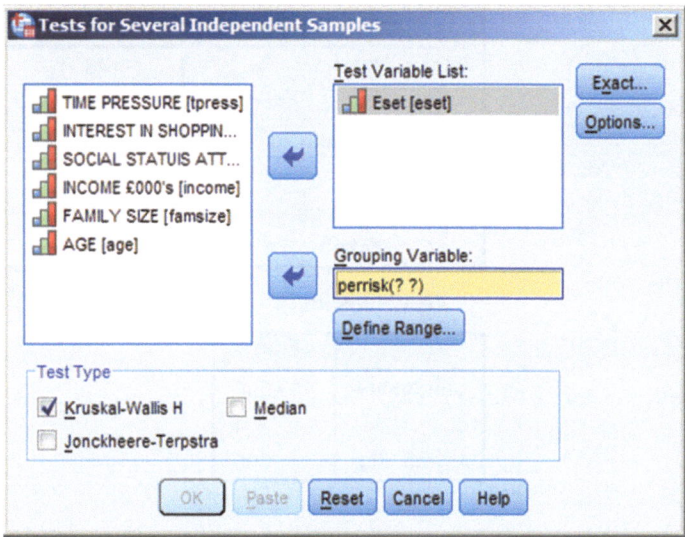

Fig. 6.7 Tests for Several Independent Samples dialogue box

The Kruskal–Wallis H test is the default of the two choices. The variable ESET is entered into the 'Test Variable List' and PERRISK is again the 'Grouping Variable'. Clicking the Define Range… button produces the *Several Independent Samples: Define Range dialogue box* of Fig. 6.8 and the 'Minimum' code is 1 and the 'Maximum' code is 5. Once more, the Options… button in Fig. 6.7 makes available descriptive statistics and quartiles. The results of applying the Kruskal–Wallis test to our consumer data are presented in the *IBM SPSS Statistics Viewer* of Fig. 6.9.

Fig. 6.8 Several Independent Samples: Define Range dialogue box

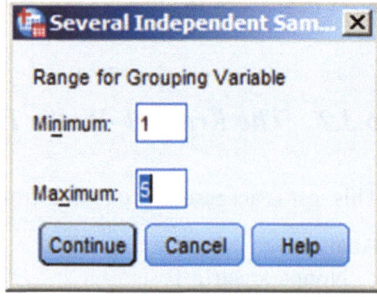

Fig. 6.9 Results of applying the Kruskal–Wallis test

Kruskal-Wallis Test

Ranks

	PERCEIVED RISK	N	Mean Rank
Eset	very low	48	33.77
	low	24	58.46
	medium	9	79.00
	high	27	91.74
	very high	42	121.74
	Total	150	

Test Statistics[a,b]

	Eset
Chi-Square	103.000
df	4
Asymp. Sig.	.000

a. Kruskal Wallis Test

b. Grouping Variable: PERCEIVED RISK

The significance level of 0.000 associated with the chi square statistic of 103.000 means that we reject H_0: the samples have been drawn from populations with the same distribution. Examination of the mean ranks once more indicates increasing evoked set sizes with increasing risk. Certainly the highest risk perceivers are markedly different from the other groups. It would be informative for the user to construct a boxplot of the evoked set sizes over our five groups of perceived risk which would reinforce this point. There is a multiple comparisons procedure associated with the Kruskal–Wallis ANOVA.

Part IV
Methods of Business Forecasting

Chapter 7
Bivariate Correlation and Regression

An essential part of managing any organisation be it governmental, commercial, industrial or social, is planning for the future by generating adequate forecasts of factors that are central to that organisation's successful operation. Methods of forecasting fall into two groups; qualitative and quantitative. Among the former fall expert judgment and intuitive approaches. Such methods are particularly used by management when conditions in the past are unlikely to hold in the future. This chapter develop the quantitative regression approach to forecasting.

In all branches of scientific enquiry, no statistical tool has received the attention given to *regression analysis* during the last 30 years. A frequent misconception among non-statistical users of regression is that the technique's prime objective is to generate forecasts. Besides being a forecasting tool, regression analysis attempts to shed light on the mechanisms that relate variables. Knowledge of such mechanisms would, in some circumstances, permit a degree of control. For example, knowledge of how certain factors contribute towards the production of defective industrial items might assist in reducing the defective rate. Knowledge of the factors that drive changes in share prices would help in portfolio selection. Regression is a tool for forecasting and explanation. (Time series analysis, a technique that is discussed later, is solely a forecasting tool.) Of all regression models, the linear model is the most widely applied.

The next two sections review respectively bivariate (two-variable) correlation and linear regression. In terms of linear regression, Pearson's product-moment correlation coefficient is the most important measure of correlation. There are assumptions that underpin the regression method and which require attention before applying the method, even in the simple bivariate case. A fourth section illustrates the application of regression and correlation in IBM SPSS Statistics. Statistical and graphical plots are illustrated.

© Springer International Publishing Switzerland 2016
A. Aljandali, *Quantitative Analysis and IBM® SPSS® Statistics*,
Statistics and Econometrics for Finance, DOI 10.1007/978-3-319-45528-0_7

7.1 Bivariate Correlation

There are three types of bivariate correlation available in IBM SPSS Statistics; *Pearson's product moment correlation coefficient* (r), *Spearman's rank correlation coefficient* (ρ) and *Kendall's tau b*. These and all other correlation measures perform different tasks. Pearson's r is a measure of linearity and is thus the most important in relation to linear regression. In the bivariate case, if two variables X_i and Y_i (i = 1, 2, ..., n where n is the number of observations) are not linearly related or correlated, there seems little point in proceeding with linear regression to generate forecasts. If the bivariate data points fall on a perfectly straight line with positive gradient, r would take the value 1; if they fall on such a line with negative gradient, r would be -1. If the Pearsonian coefficient is close to zero, it may well indicate an almost random scatter of the points. However, r being close to zero only indicates the absence of linearity. The variables at hand may exhibit a marked curved relationship. A U-shaped quadratic relationship would generate a value of r close to zero. Certainly in the bivariate case, it is wise to construct a scatterplot in IBM SPSS Statistics of the variables, to ensure that non-linear relationships are not ignored.

A common precursor to employing bivariate regression is to test whether or not the gathered sample data have been drawn from a population in which the Pearsonian correlation coefficient R is zero. As already stated, if the data have been drawn from such a population, then linear regression would seem a pointless exercise. The statistical test that:

H_0: sample is drawn from a population in which $R = 0$

is part of the IBM SPSS Statistics Correlations routine. The appropriate test statistic under H_0 is:

$$ r \frac{\sqrt{n-2}}{1-r^2} $$

which is distributed as the t statistic with $n-2$ degrees of freedom; n is the number of sampled bivariate points. For example, suppose we were testing a one tailed alternative H_1: $R > 0$, then if n = 47 readings produced a sample correlation coefficient of r = 0.34, then the above equation would produce a test statistic with numerical value 2.4253. From statistical tables $P(t_{45} > 2.0141) = 0.05$, so we reject the hypothesis that $R = 0$. IBM SPSS Statistics prints exact significance levels for this test statistic.

The square of the Pearsonian coefficient (usually expressed as a percentage) is called the *coefficient of determination* $r^2\%$, so if r = 0.8 then $r^2\% = 64\%$. The coefficient of determination measures the amount of variation in the Y (the dependent variable) that is 'explained' by a linear relationship with the X (the independent variable). This parameter is a better indicator of the extent of linearity present than is r. For example, a sample correlation coefficient of r = 0.7 might seem relatively high. However, when viewed in the context of the coefficient of determination, we see that less than half of the variation in the Y is explained by a linear relationship with the X. The 51 % of variation in the Y variable that is unexplained could be due to errors in measurement and to variables that have not been included in the study. The latter variables are called *extraneous factors* and would be of consequence in the above example.

Spearman's rank correlation by its very title is a bivariate measure of association when the data comprise ranks, for example, two consumers might rank five compet-ing brands from 1 ("most preferred") to 5 ("least preferred"). Spearman's ρ mea-sures the degree of similarity between the consumers' rank scores. In fact, it may be shown that the numerical value of Spearman's ρ is the same as the value that would be obtained if the ranked scores were inserted into the formula for Pearson's r. Spearman's ρ is thus a measure of the degree of linearity between the ranked scores. If the two consumers agreed exactly in their rankings of the brands, then these ranks would be perfectly linearly related and $r = 1$. If the ranks were the exact opposite of each other, then $r = -1$.

Just as in the cases of Spearman's coefficient, Kendall's tau b is a measure of association that applies to ordinal data and may take values between ±1 inclusive. Essentially, the latter statistic looks at all possible pairs of cases and examines whether or not the ranks are in the same order.

7.2   Linear Least Squares Regression for Bivariate Data

If the coefficient of determination indicates that there is adequate linearity present in the gathered bivariate data set (e.g. $r^2 > 80\%$), then we may proceed to fit a line through the data points and use this line for forecasting purposes. This line is called *the linear least squares regression line*, denoted in the bivariate case by $\overset{\frown}{Y} = a + bX$, where $\overset{\frown}{Y}$ represents the forecasted or predicted values. The method for determining the optimal value for the parameters a and b involves minimizing the sum of squares of the errors (or *residuals*) about the regression line i.e.:

$$\min \sum \left(Y - \overset{\frown}{Y} \right)^2$$

where the expression $(Y - \overset{\frown}{Y})$ is the difference between the observed and predicted values of Y, namely the *error* or *residual*. We are unable to minimise the sum of the residuals themselves, because this sum is always zero. Gathered data are most often a sample from a particular population. The population regression line is denoted by $Y = \alpha + \beta X$. Given the importance of gradients in fields such as Economics (in that gradients reflect rates of change; for example), it is often desirable to make inferences about β based on the value of b derived from the sample data. In particu-lar, it is most frequently hoped that we have sufficient sample evidence to reject:

$$H_0 : \beta = 0.$$

For example, suppose it is believed that revenue (Y) in sales regions depends on the number of television advertisements (X) shown in those regions over a particular period. If we failed to reject the above null hypothesis, this would infer that the population gradient is zero, i.e. for every advertisement shown, there is no increase in sales revenue.

The above null hypothesis is assessed via the test statistic:

$$(b-\beta)/(\text{SE of } b),$$

where SE represents *standard error*. Recalling that statistical tests are conducted under the assumption that the null hypothesis is true, the value of β in the above would be zero. However, if one wished to test H_0: $\beta = 4$ (e.g. an accountant may wish the rate of return on a project of at least four units per financial unit invested), then β would take a value of four in the formula for the test statistic. The above test statistic is distributed as the t statistic with $n-2$ degrees of freedom. The method of calculating the standard error of b is tedious and requires some statistical expertise, so interested readers are referred to standard statistical texts.

7.3   Assumptions Underlying Linear Least Squares Regression

There are a series of statistical assumptions that underpin the least squares regression model and which will influence our selections from the IBM SPSS Statistics dialogue boxes illustrated in the next subsection. Most of the assumptions refer to the residuals or errors from the regression analysis. Remember that a residual is simply the difference between the observed value of the dependent variable at a particular value of X and the predicted value of the dependent variable at that value of X.

The assumptions that underlie regression are:

- the relationship between the Y and X should be adequately approximated by a straight line,
- the residuals should have zero mean,
- the residuals should have constant variance (called the *homoscedasticity* assumption),
- the residuals should not be correlated and
- the residuals should be normally distributed.

A scatter diagram will help to indicate the adequacy or otherwise of using a linear regression approach. In the bivariate case, the coefficient of determination gives a less subjective approach. The residuals will have a zero mean due to the mechanics whereby the least squares line is computed.

If the homoscedasticity assumption is violated, the constants a and b (called *regression coefficients*) in the regression equation are still unbiased estimates of their population equivalents α and β, but they are no longer minimum variance estimators. Put more simply, if the variance or spread of the errors is, for example, increasing as we move along the regression line, then why are we fitting a line to the data? Certainly, forecasts into the future are likely to possess high error in this situation. A simple way of assessing the constant residual variance assumption is graphical. A plot of the residuals against the predicted values $\overset{\frown}{Y}$ (i.e. the regression line) should show no evidence of varying spread.

Correlation between the residuals is common when we have economic data recorded over time. This phenomenon is called *temporal autocorrelation* and again may be assessed graphically. A plot of the residuals over time should evidence a random pattern i.e. the residuals are independent of time. Again, the regression coefficients are still unbiased but they are no longer minimum variance estimates in the face of autocorrelation. Confidence intervals for α and β tend to be narrower than they should be. (There is a formal test for autocorrelation called the *Durbin–Watson test* and which is available in IBM SPSS Statistics.)

It should be noted that small departures from normality of the residuals does not affect the linear regression model greatly. However, gross violations of this assumption can seriously distort the test statistic for the population gradient β and associated confidence intervals. The normality of the residuals may be assessed via a normal probability. Should there be sufficient residuals, then a histogram will give a graphical assessment of this assumption. It is also possible to save the residuals in the existing IBM SPSS Statistics data file and subject them to the Shapiro–Wilks available in the Explore procedure (Sect. 3.2).

This subsection has indicated that there are various plots that may assist in the assessment of the regression assumptions and naturally, these plots are available in IBM SPSS Statistics. By default, such plots *standardize* the variables in question, such as the residuals. This is to remove the problem of the unit of measurement. For example, if we had a residual or error of $50,000,000 in a problem, this may seem immense. However, if the problem involved forecasting the gross national product of the U.S.A., then the residual is far from serious! Standardization of residuals overcomes the problem of units. Standardization leads the residuals to have a mean of zero and variance of one.

7.4   Bivariate Correlation and Regression in IBM SPSS Statistics

The file EARNINGS.SAV contains the returns on assets for 30 American firms in terms of several potential explanatory variables. These variables along with their IBM SPSS Statistics names are:

- CAPGWTH—capital spending growth,
- INDRET—industry return on assets,
- INDSALES—industry sales growth,

- INVTURN—inventory turnover,
- OPINCGTH—operating income growth,
- OPINCRAT—the ratio of operating income to total assets and
- SALESGTH—sales growth.

The first stage in the presented analysis of this section is to see which of these seven variables has the strongest linear relationship with the variable to be explained, COASSETS. This will be achieved by running the IBM SPSS Statistics bivariate correlations procedure.

To access the *Bivariate Correlations dialogue box* of Fig. 7.1, from the Data Editor click:

Analyze
  Correlate
    Bivariate…

and you will see the availability of the three bivariate correlation measures discussed in Sect. 7.1, with Pearson being the default. A two-tailed test of significance (the default) has been selected in this case. This option should be used when the user is not in a position to know in advance the direction (positive or negative correlation) of the relationship.

The dialogue box of Fig. 7.1 in fact generates the value of Pearson's r for all pairs of the eight variables in the box titled 'Variables'. Figure 7.2 presents the statistical output. The variable COASSETS ('FIRM'S RETURN ON ASSETS') is regarded

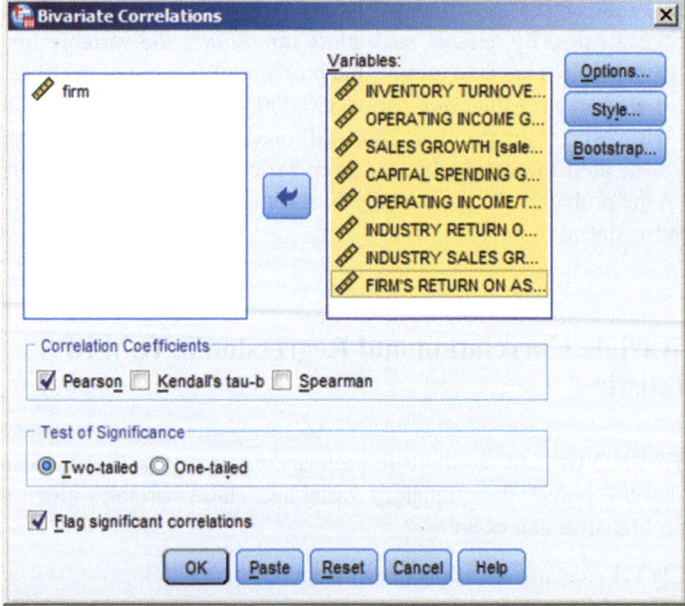

Fig. 7.1 The Bivariate Correlations dialogue box

Correlations

		INVENTORY TURNOVER	OPERATING INCOME GROWTH	SALES GROWTH	CAPITAL SPENDING GROWTH	OPERATING INCOME/TOT AL ASSETS	INDUSTRY RETURN ON ASSETS	INDUSTRY SALES GROWTH	FIRM'S RETURN ON ASSETS
INVENTORY TURNOVER	Pearson Correlation	1	-.213	-.484**	-.270	-.389*	-.608**	-.752**	-.515**
	Sig. (2-tailed)		.257	.007	.149	.034	.000	.000	.004
	N	30	30	30	30	30	30	30	30
OPERATING INCOME GROWTH	Pearson Correlation	-.213	1	.443*	.138	-.106	.091	.310	.047
	Sig. (2-tailed)	.257		.014	.468	.578	.634	.096	.804
	N	30	30	30	30	30	30	30	30
SALES GROWTH	Pearson Correlation	-.484**	.443*	1	.334	.351	.289	.453*	.250
	Sig. (2-tailed)	.007	.014		.072	.057	.122	.012	.183
	N	30	30	30	30	30	30	30	30
CAPITAL SPENDING GROWTH	Pearson Correlation	-.270	.138	.334	1	.282	.154	.189	.311
	Sig. (2-tailed)	.149	.468	.072		.132	.416	.318	.094
	N	30	30	30	30	30	30	30	30
OPERATING INCOME/TOTAL ASSETS	Pearson Correlation	-.389*	-.106	.351	.282	1	.347	.431*	.763**
	Sig. (2-tailed)	.034	.578	.057	.132		.060	.017	.000
	N	30	30	30	30	30	30	30	30
INDUSTRY RETURN ON ASSETS	Pearson Correlation	-.608**	.091	.289	.154	.347	1	.793**	.380*
	Sig. (2-tailed)	.000	.634	.122	.416	.060		.000	.039
	N	30	30	30	30	30	30	30	30
INDUSTRY SALES GROWTH	Pearson Correlation	-.752**	.310	.453*	.189	.431*	.793**	1	.410*
	Sig. (2-tailed)	.000	.096	.012	.318	.017	.000		.025
	N	30	30	30	30	30	30	30	30
FIRM'S RETURN ON ASSETS	Pearson Correlation	-.515**	.047	.250	.311	.763**	.380*	.410*	1
	Sig. (2-tailed)	.004	.804	.183	.094	.000	.039	.025	
	N	30	30	30	30	30	30	30	30

**. Correlation is significant at the 0.01 level (2-tailed).

*. Correlation is significant at the 0.05 level (2-tailed).

Fig. 7.2 Output from running bivariate correlation

as the dependent variable. Figure 7.2 shows that COASSETS is significantly correlated with the INVTURN. The value of the Pearsonian correlation between COASSETS and INVTURN is −0.515 for the 30 sample observations. The significance associated with the null hypothesis:

H_0: the population correlation coefficient, R, between these two variables is zero

is p=0.004. The null hypothesis is, therefore, rejected and the correlation is significantly different from zero. The variable COASSETS is also significantly correlated with INDSALES, as indicated by the levels of significance being less than 0.025 for these two tailed tests. All levels of significance are based on computation of the test statistic and the t distribution with n−2 degrees of freedom.

The variable OPINCRAT, in fact, has the most significant correlation with COASSETS and will, therefore, be used to illustrate bivariate linear regression. The correlation exercise, therefore, suggests that the variable OPINCRAT is the most important determinant of the values of the variable COASSETS.

From the Data Editor, click:

Analyze
   Regression
      Linear…

to obtain the *Linear Regression dialogue box* of Fig. 7.3. The 'Dependent' variable in this example is COASSETS which is entered into the associated box in the usual manner. The 'Independent' variable is OPINCRAT. There are several methods for conducting regression analysis, most of which are pertinent if the researcher is

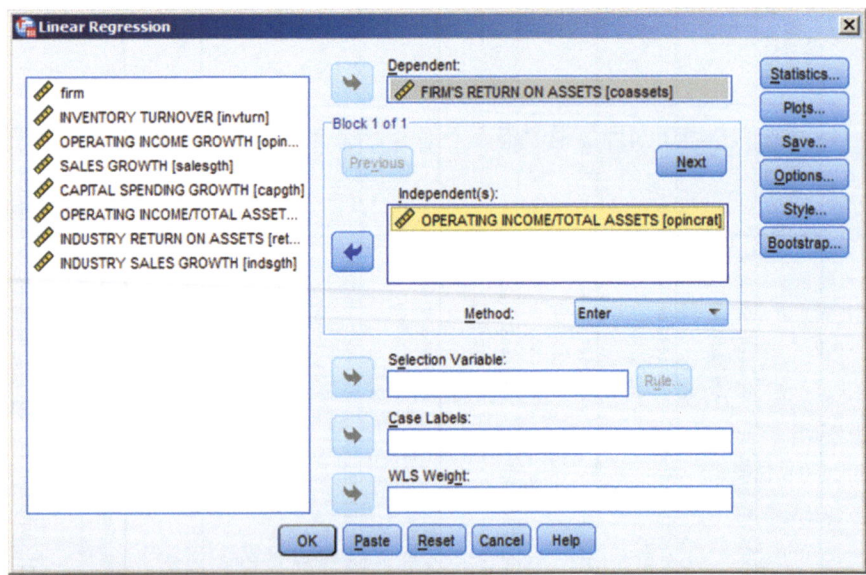

Fig. 7.3 The Linear Regression dialogue box

Fig. 7.4 The Linear
Regression: Statistics
dialogue box

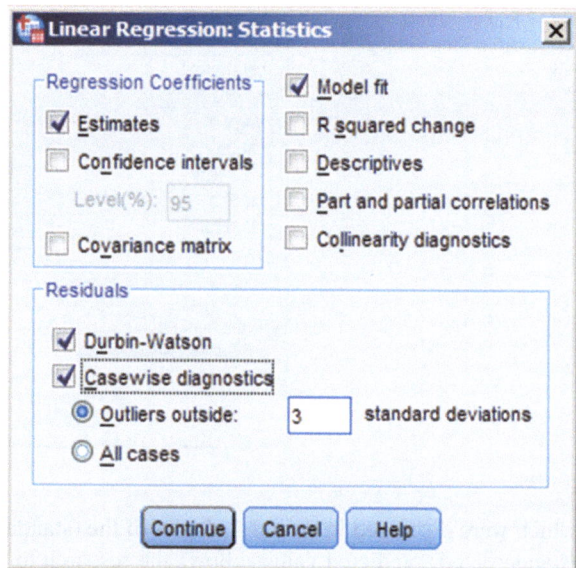

pursuing a multivariate analysis. Suffice it to say at present that the default procedure in the 'Method' box of Fig. 7.3 is called the "Enter Method" and will be chosen
here. Generally, this method enters all of the independent variables in one step.
Here, of course, we only have one independent variable.

Click the Statistics… button at the top right hand corner of the *Linear Regression
dialogue box* to obtain the *Linear Regression: Statistics dialogue box* of Fig. 7.4. By
default, estimates of the regression coefficients are produced. Confidence intervals
for these coefficients are optional as are various descriptive statistics. Note that the
Durbin–Watson test for autocorrelation is selected from this dialogue box if desired,
but our COASSETS data are not temporal, so this test is irrelevant. *Casewise diagnostics* (i.e. firm by firm) of standardized residuals for all cases has been chosen
from this dialogue box.

Note that under the heading 'residuals', this dialogue box accommodates the
detection of 'outliers'. Loosely speaking, *outliers* are points that are far distant from
the regression line i.e. they have large positive or negative residuals. They could
represent data input errors. They could also be points of special interest that merit
further study or separate analysis. Recall that IBM SPSS Statistics standardizes the
residuals (mean of zero and variance of one). By default, points more than three
standard deviations either side of the regression line are regarded as outliers in IBM
SPSS Statistics. The user may change this limit in the dialogue box of Fig. 7.4.
Click the Continue button to return to the *Linear Regression dialogue box* of
Fig. 7.3.

At the bottom of the *Linear Regression dialogue box* is the Plots… button that
gives rise to the *Linear Regression: Plots dialogue box* of Fig. 7.5, which permits
graphical evaluations of the assumptions underlying the regression method and

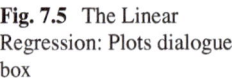

Fig. 7.5 The Linear
Regression: Plots dialogue
box

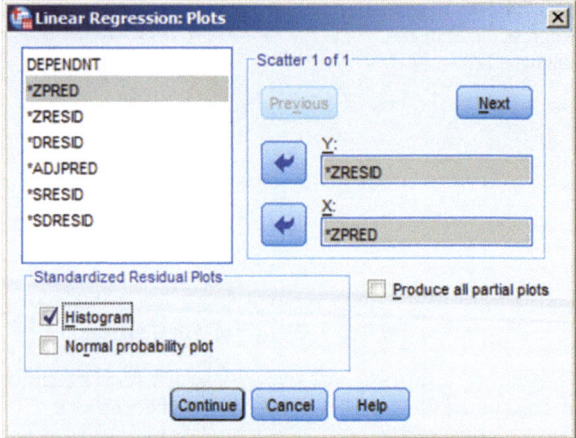

which were discussed in Sect. 7.2. A plot of the (standardized) residuals against the (standardized) predicted values allows the research to judge if homoscedasticity is present. In IBM SPSS Statistics, the standardized residuals are denoted by *ZRESID and the standardized predicted values by *ZPRED. These are respectively clicked into the boxes labelled $\underline{Y}$ and $\underline{X}$, as shown, via the arrow buttons. In Fig. 7.5, a histogram has also been selected to assess the normality assumption pertaining to the residuals.

Click the Continue button to return to the *Linear Regression dialogue box* of Fig. 7.3. Again at the top right of this dialogue box is the Save... button which accesses the *Linear Regression: Save dialogue box* of Fig. 7.6. Many of the options here require advanced knowledge of regression techniques. However, the user may wish to save 'Unstandardized' and 'Standardized' predicted and residual values for further study or graphical analysis. The appropriate boxes are simply clicked and a cross appears in each upon selection. Click the Continue button to return to the *Linear Regression dialogue box* and then the OK button to perform the regression analysis. Figure 7.7 presents part of the results of this regression analysis in the IBM SPSS Statistics Viewer.

The value of the coefficient of determination is $r^2 \% = 58.2\%$, so nearly 40 % of the variation in company COASSETS remains unexplained after the introduction of the operating income/total assets ratio. Clearly, some of the other variables that are significantly correlated with these firms' returns should be introduced into our analysis, creating a multivariate regression problem. The value of the Pearsonian correlation between the variables COASSETS and OPINCRAT is 0.763 (p=0.000).

The equation of least squares linear regression is:

$$COASSETS = 11.697 + .639 * (OPINCRAT),$$

but this bivariate equation of regression would, in all probability, be inadequate for forecasting purposes in that r^2 is not sufficiently high. The above figure permits

Fig. 7.6 The Linear Regression: Save dialogue box

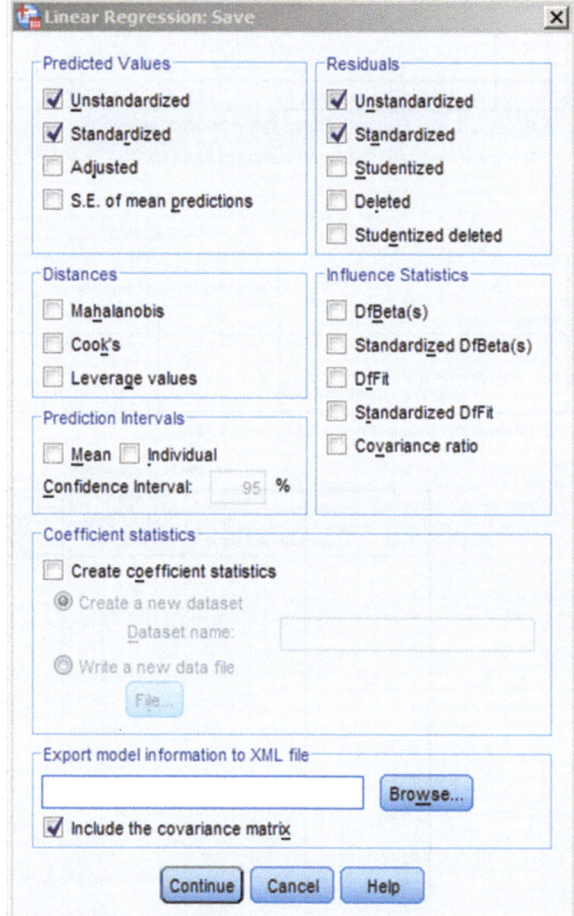

study of the hypothesis that the population regression line has a zero gradient (i.e. $H_0: \beta = 0$). From Fig. 7.7, we, therefore, derive a test statistic of:

$$b / (\text{SE of } b) = 0.639 / 0.102 = 6.264 \,(\text{remember } \beta = 0 \text{ under the null hypothesis}),$$

which is distributed as a t statistic with $n - 2 = 28$ degrees of freedom. This test statistic is part of the output of Fig. 7.7 and has a significance level of $p = 0.000$ to three decimal places. We thus reject the null hypothesis and conclude that the population gradient is non-zero. Our best estimate of β is simply the sample value of 0.639. This means that on average, a one unit increase in the operating income to total assets ratio generates a 0.639 increase in company returns. A 95 % confidence interval for the population gradient is in fact given by:

$$P(0.429 < \beta < 0.848) = 0.95$$

Model Summary[b]

Model	R	R Square	Adjusted R Square	Std. Error of the Estimate	Durbin-Watson
1	.763[a]	.582	.567	4.3907	2.196

a. Predictors: (Constant), OPERATING INCOME/TOTAL ASSETS

b. Dependent Variable: FIRM'S RETURN ON ASSETS

Coefficients[a]

Model		Unstandardized Coefficients		Standardized Coefficients	t	Sig.
		B	Std. Error	Beta		
1	(Constant)	11.697	3.700		3.161	.004
	OPERATING INCOME/TOTAL ASSETS	.639	.102	.763	6.243	.000

a. Dependent Variable: FIRM'S RETURN ON ASSETS

Casewise Diagnostics[a]

Case Number	Std. Residual	FIRM'S RETURN ON ASSETS	Predicted Value	Residual
1	-.387	31.9	33.601	-1.7008
2	.741	42.6	39.348	3.2519
3	1.348	42.2	36.283	5.9172
4	.624	35.0	32.260	2.7403
5	-1.441	24.4	30.727	-6.3271
6	.156	36.2	35.517	.6835
7	.995	36.5	32.132	4.3680
8	.530	33.5	31.174	2.3259
9	.307	34.5	33.154	1.3462
10	-.388	30.3	32.004	-1.7043
11	-2.063	38.4	47.458	-9.0582
12	.292	31.5	30.216	1.2838
13	.736	30.7	27.470	3.2297
14	.997	44.3	39.923	4.3772
15	-.197	32.8	33.665	-.8646
16	-1.934	17.7	26.193	-8.4931
17	-.738	31.0	34.239	-3.2394
18	.117	35.2	34.686	.5136
19	.626	38.2	35.453	2.7473
20	-.402	28.9	30.663	-1.7632
21	-.313	32.1	33.473	-1.3730
22	1.323	44.9	39.093	5.8073
23	-.924	38.1	42.158	-4.0579
24	.228	24.0	23.000	.9999
25	-.977	33.4	37.688	-4.2878
26	1.654	46.8	39.540	7.2603
27	-.332	27.1	28.556	-1.4559
28	.668	34.3	31.366	2.9343
29	-1.835	28.1	36.155	-8.0551
30	.591	42.9	40.306	2.5940

a. Dependent Variable: FIRM'S RETURN ON ASSETS

Fig. 7.7 Part of the output from running bivariate regression

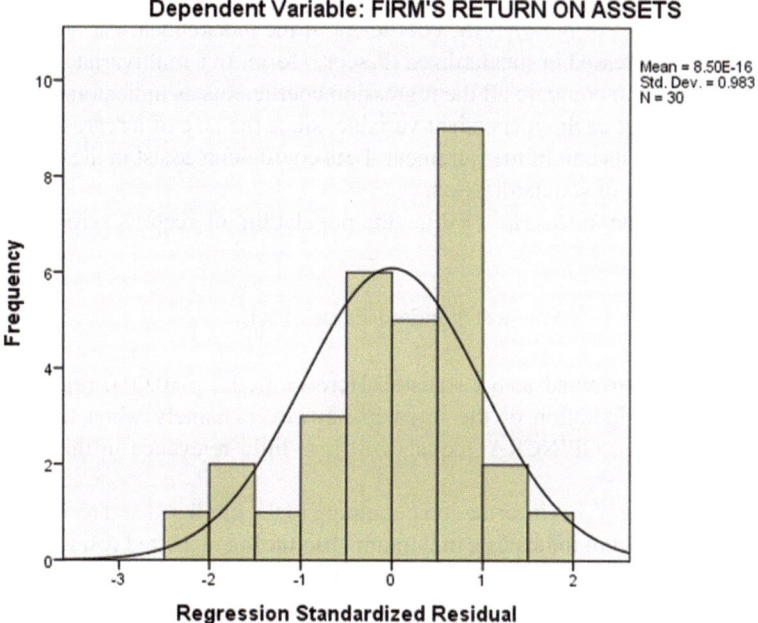

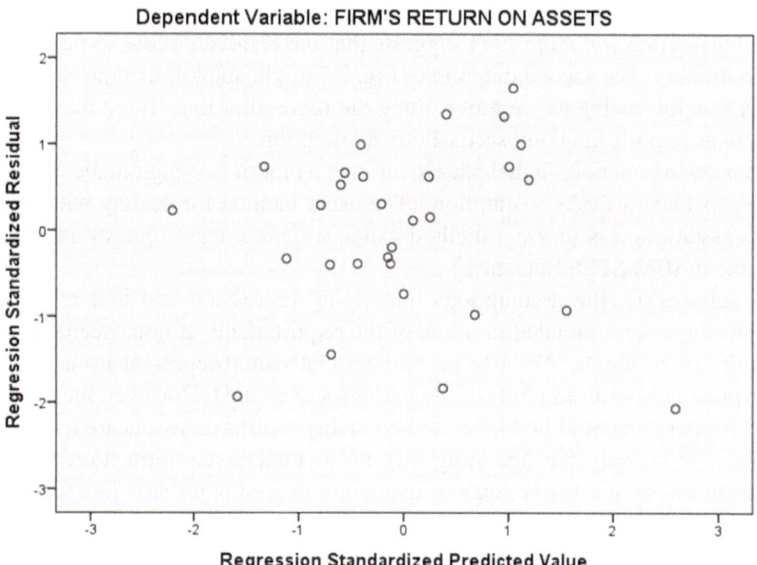

Fig. 7.7 (continued)

and note that a value of $\beta=0$ is not contained in this interval, as is expected after conducting the hypothesis test on the population gradient. The *beta coefficient* (=0.763) reported in Fig. 7.7 is the coefficient of the independent variable when all variables are expressed in standardized (Z score) form. In a multivariate problem, it would be wrong to compare all the regression coefficients as indicators of the relative importance of each independent variable, since the size of a regression coefficient depends on its unit of measurement. Beta coefficients assist in the comparison process by means of standardization.

(It also possible to test H_0: $\alpha=0$ i.e. the population intercept is zero via the test statistic:

$$a\,/\left(\text{Standard Error of }a\right),$$

which is also distributed as a t statistic. Here t=3.161, p=0.004, reject the null hypothesis. Consideration of the intercept, however, namely when the value of COASSETS when OPINCRAT equals zero has little relevance in this particular context.)

Included in Fig. 7.7 is information pertaining to the predicted and residual values obtained. This Figure indicates a maximum absolute standardized residual of 2.362 associated with company number 14. If we consider three standard deviations away from the regression line as the characteristic of an outlier, our study has thrown up no outliers. However, some researchers consider cases that are over two standard deviations away from the regression line as outliers, in which instance companies numbered 14 and 20 would be so considered.

The histogram (*on page 131*) suggests that the residuals show some departures from normality. The second diagram of Fig. 7.7 might suggest that the spread of the residuals is increasing as we move along the regression line. There may well be a problem as regards the homoscedasticity assumption.

An outward-opening funnel pattern on such a plot of is symptomatic of violation of the constant variance assumption. (The usual method for dealing with violation of this assumption is to use a method called *weighted least squares* and which is available in IBM SPSS Statistics.)

To summarize, the assumptions underlying regression and that relate to the bivariate case seem violated in terms of the requirements of homoscedasticity and normality of residuals. We have no outliers that would represent firms exhibiting non-typical behaviour in terms of the variables examined. However, the coefficient of determination should be higher and company returns on assets are inadequately explained by simply the operating income to total assets ratio. Forecasting the returns on assets of other companies using our derived regression line would probably be prone to unacceptable error. We, therefore, need to treat the analysis in a multivariate manner and introduce other, salient independent variable(s).

Chapter 8
Elementary Time Series Methods

Much of the data used and reported in Economics is recorded over time. The term *time series* is given to a sequence of data, (usually intercorrelated), each of which is associated with a moment in time. Example like daily stock prices, weekly inventory levels or monthly unemployment figures are called discrete series, i.e. readings are taken at set times, usually equally spaced. The form of the data for a time series is, therefore, a single list of readings taken at regular intervals. It is this type of data that will concern us in this chapter.

There are two aspects to the study of time series. Firstly, *the analysis phase* attempts to summarize the properties of a series and to characterize its salient features. Essentially, this involves examination of a variable's past behaviour. Secondly, the *modeling phase* is performed in order to generate future forecasts. It should be noted that in time series, there is no attempt to relate the variable under study to other variables. This is the goal of regression methods. Rather, in time series analysis, movements in the study variable are 'explained' only in terms of its own past or by its position in relation to time. Forecasts are then made by extrapolation.

IBM SPSS Statistics has available several methods of time series analysis. This chapter describes two of the more simple time series methods, *seasonal decomposition* and *one parameter exponential smoothing*. Suffice it to say that these methods involve much tedious arithmetic computation and may only realistically be performed on a computer. The first section of this chapter reviews the logic of seasonal decomposition.

Graphics are particularly useful in time series studies. They may, for example, highlight regular movements in data and which may assist model specification or selection. Given the excellent graphics capabilities of IBM SPSS Statistics, the package is particularly amenable to time series analysis. The generation of various plots of temporal data over time is assisted if date variables are defined in IBM SPSS Statistics. Indeed, seasonal decomposition requires their definition and the method for achieving this is described in the second section of this chapter.

© Springer International Publishing Switzerland 2016
A. Aljandali, *Quantitative Analysis and IBM® SPSS® Statistics*,
Statistics and Econometrics for Finance, DOI 10.1007/978-3-319-45528-0_8

There then follow two sections that describe two types of widely used decomposition methods—the *additive* and *multiplicative* models. Both are illustrated and the terminology used by IBM SPSS Statistics is defined. Next follows a review of exponential smoothing, again followed by an illustration in IBM SPSS Statistics.

8.1   A Review of the Decomposition Method

A major aspect of selecting appropriate time series models is to identify the basic patterns or components inherent in the gathered data. Time series data consist of some or all of the following components:

- a trend (T), which is a persistent, long run, upward or downward movement in the data,
- seasonal variation (S), which occurs during the year and then is repeated on a yearly basis for example, sales of jewellery in the United States peak in December,
- a cycle (C), which is represented by relatively slow wave-like fluctuations about the trend in the behaviour of the series. A cycle is measured from peak to peak or trough to trough and
- irregular fluctuations (I), which are erratic movements in the data over time. They are usually due to unpredictable, outside influences, such as industrial strikes.

The decomposition method of time series analysis assumes that the data may be broken down into these components. There are two types of time series model—the *additive* and the *multiplicative*. If we denote the variable under examination by Y- is the sum of the aforementioned four components:

$$Y_t = T + S + C + I$$

where the subscript t represents time period t. If one of the components is absent, then its value is zero. This model assumes that the magnitude of the seasonal movement is constant over time, as shown in Fig. 8.1. The multiplicative model is of the form:

$$Y_t = T \cdot S \cdot C \cdot I$$

This model assumes that the magnitude of the seasonal movement increases or decreases with the trend, as shown in Fig. 8.2. The multiplicative model is generally the more relied upon, in that it identifies the integral components of many real economic time series.

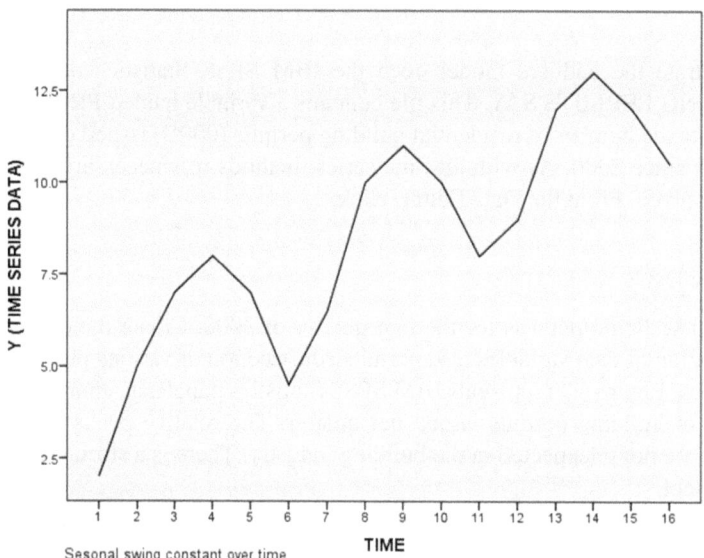

Fig. 8.1 An Additive Time Series Model

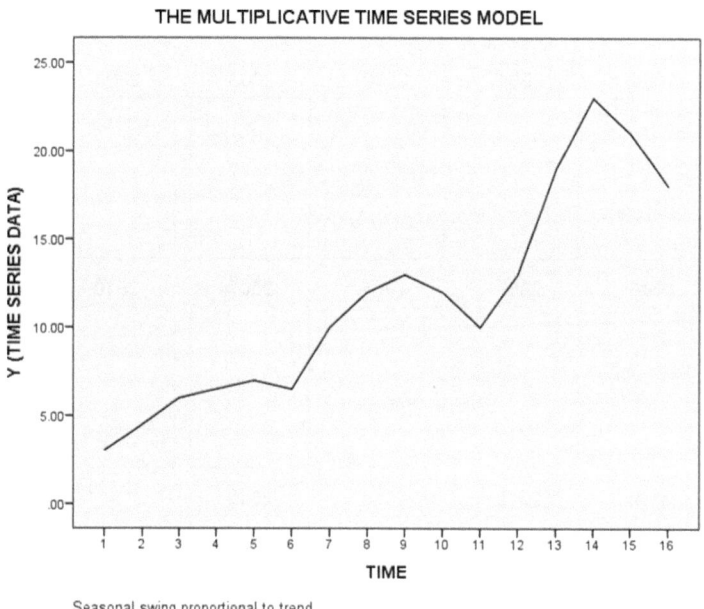

Fig. 8.2 A Multiplicative Time Series Model

8.2   The Additive Model of Seasonal Decomposition

To illustrate the additive model open the IBM SPSS Statistics data file called
BUILDING PERMITS.SAV. This file contains a variable named PERMITS which
represents the number of residential building permits (000's) issued each quarter in
the USA since 2000. As with all-time series methods it is necessary to define the
dates involved. From the Data Editor, click:

Data
   Define Dates…

The first datum point is for the first quarter of 2000. Create the data variables.
The creation of data variables also permits construction of various plots of the time
series data. Figure 8.3 represents IBM SPSS Statistics panel data charts showing the
number of building permits issued per quarter. The relative peaks at the second
quarters are not unexpected in the building industry. There is a strong seasonal pattern present.

Integral to the application of the additive model is the concept of *centred moving
average (CMA)*. The point behind a moving average is that seasonal (S) and irregular (or error) fluctuations (I) in time series data make it difficult to discern trend (T)
and cyclical patterns (C) in the data. In general, all types of moving average *smooth*

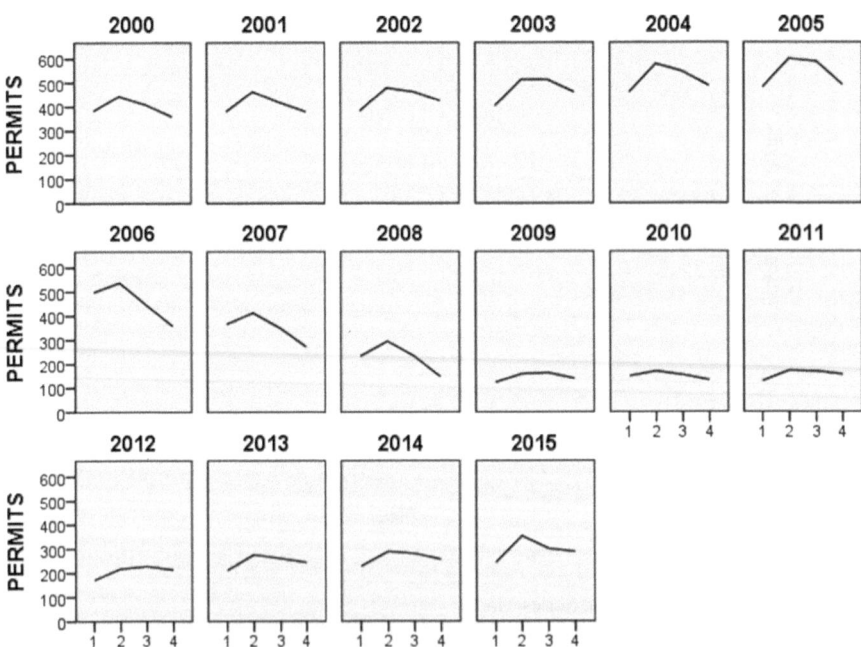

Fig. 8.3 Number of issued building permits per quarter panelled per year

the data and *eliminate the seasonal and irregular variation whilst retaining the trend and cyclical variation.*

When the number of time periods is even, the approach illustrated below is used in the computation of the CMA's. Table 8.1 reproduces part of our data file BUILDING PERMITS.SAV and will now be used to show how to compute CMA's based on an even number of time periods, as well as the method of seasonal decomposition itself. Given that the building permit data are quarterly and that Fig. 8.3 has stressed the seasonality inherent in the data, we will compute CMA's based on $k=4$ time periods.

Consider the first four readings in Table 8.2. Based on $k=4$ time periods, we compute the four period moving averages as:

$$(382.812 + 442.842 + 407.953 + 358.660) / 4 = 1592.267 / 4 = 398.067$$

This moving average figure is written between the second and third quarter of 2000. Consider the next block of four figures, the second moving average is:

$$(442.842 + 407.953 + 358.660 + 378.991) / 4 = 1588.446 / 4 = 397.111,$$

Table 8.1 Smoothing of quarterly data

Date	No. of permits	Moving total	CMA
2000:Q1	382.812		
2000:Q2	442.842	398.067	
2000:Q3	407.953	397.11	397.589
2000:Q4	358.660	401.963	399.537
2001:Q1	378.991	404.270	403.116
2001:Q2	462.249	409.169	406.719
2001:Q3	417.180	409.766	409.468
2001:Q4	378.256		

Table 8.2 Deseasonalizing time series data under the additive model

Date	No. of permits	CMA	S+I	S	D
2000:Q1	382.812			−30.93	413.742
2000:Q2	442.842			41.17	401.672
2000:Q3	407.953	397.589	10.364	16.96	390.993
2000:Q4	358.660	399.537	−40.877	−27.20	385.86
2001:Q1	378.991	403.116	−24.125	−30.93	409.921
2001:Q2	462.249	406.719	55.53	41.17	421.079
2001:Q3	417.180	409.468	7.712	16.96	400.22
2001:Q4	378.256	411.771	−33.515	−27.20	405.456

which is written between the third and fourth quarter of 2000. The CMA based on $k=4$ and centred on the third quarter of 2000 is defined as the average of these two figures, namely:

$$CMA = (398.067 + 397.111)/2 = 397.589$$

similarly, the CMA based on $k=4$ and centred on the fourth quarter of 1980 is:

$$CMA = (397.111 + 401.963)/2 = 399.537$$

Centred moving averages cannot be computed for the first two quarters. Use of the derived CMA's presented in Table 8.2 permits completion of the method of seasonal decomposition. Recall that the additive model of seasonal decomposition is:

$$Y_t = T + S + C + I$$

Figure 8.4 present a plot of the raw data along with the CMA's as computed in Table 8.2. Note how the original time series has been smoothed. Seasonal (S) and

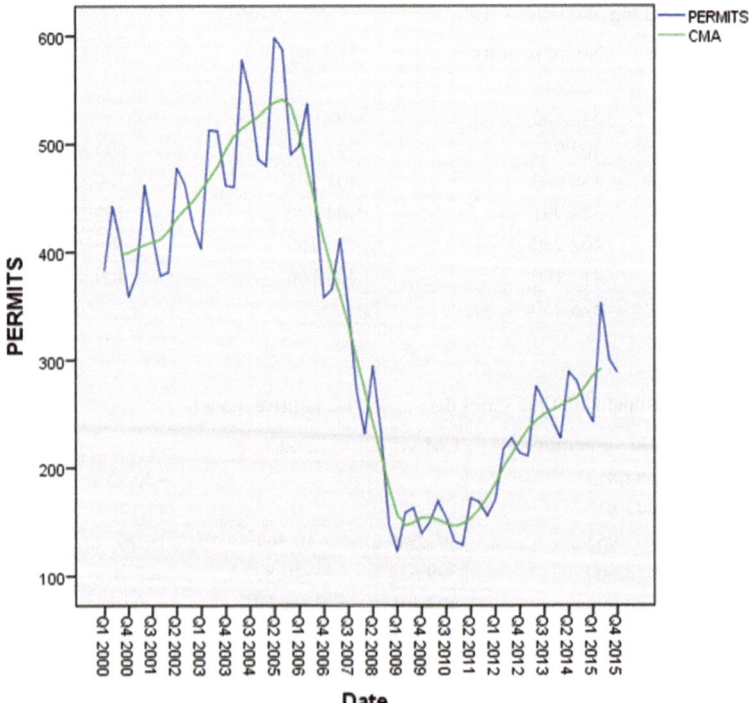

Fig. 8.4 Raw and centred moving average data

Table 8.3 Derivation of the seasonal factors for an additive model

	Q1	Q2	Q3	Q4
2000			10.364	−40.877
2001	−24.126	55.530	7.712	−33.515
2002	−38.059	47.275	22.843	−21.175
2003	−54.006	45.010	32.621	−33.462
2004	−47.163	63.630	26.049	−38.811
2005	−53.063	59.923	46.108	−45.482
2006	−12.642	60.952	2.592	−52.981
2007	−17.837	52.110	16.210	−30.120
2008	−40.526	52.465	20.378	−35.628
2009	−34.156	11.980	13.821	−14.777
2010	−4.382	17.160	5.220	−14.180
2011	−19.947	18.749	7.110	−16.898
2012	−14.730	17.537	15.276	−11.248
2013	−25.073	31.795	10.644	−9.608
2014	−30.284	27.149	15.203	−18.525
2015	−43.172	61.009		
Average	−30.611	41.485	17.271	−26.886
Sum of averages = 1.258				
Mean of this sum = (1.258/4) = 0.314				
Seasonal factors	−30.925	41.170	16.956	−27.201

irregular (I) variations have been removed leaving just the trend (T) and cyclical variation (C) which leads us to say that:

$$CMA = T + C$$

Therefore, if we subtract the computed CMA from Y_t we are left with S + I as shown in Table 8.2.

We next list the S + I according to their respective quarters as per Table 8.3. Assuming that the average of the irregular fluctuations over the quarters is zero, the mean of each column in Table 8.3 is an estimate of the seasonal (quarterly) components in our additive model. It is neater if these estimates sum to zero. They do not here (sum is 1.26). We adjust these estimates (called the *process of normalisation*) so that they do sum to zero. This is achieved by subtracting the mean of this sum (1.26/4 = 0.314) from our estimates. This yields the seasonal factors at the bottom of Table 8.3 and inserted into the fifth column of Table 8.2.

Essentially, a seasonal factor indicates the net effect of each period in a season on the level of the time series. (They are sometimes referred to as *the seasonal adjustment factors*). For example, the seasonal factor of −30.611 for the first quarter infers that the number of permits issued in this quarter is depressed by −30.611 units due to the season. Therefore, if 30.611 is added to all the first quarter figures, we arrive at what is called the *deseasonalised* or *seasonally adjusted* time series data (represented by D in Table 8.2).

Similarly, the second quarter data are inflated by 41.485 units due to the season. This figure is subtracted off all second quarter results to obtain the deseasonalised data for that quarter. This process is continued for all quarters. The point is that seasonally adjusted data have the seasonal component removed and makes it easier to study and trend and cyclical fluctuations. The researcher is able to assess whether series values have increased or decreased net of the seasonal effect. As is shown in the next subsection, these seasonal factors are used in the forecasting process.

Before performing seasonal decomposition in IBM SPSS Statistics, the data file BUILDING PERMITS.SAV must be open and the dates defined. To access the additive seasonal decomposition method, from the Data Editor click:

Analyze
  Forecasting
    Seasonal Decomposition...

which opens the *Seasonal Decomposition dialogue box* of Fig. 8.5. The IBM SPSS Statistics variable of study is PERMITS, which is entered into the box titles Variable(s). The default is the Multiplicative model, but here we are using an Additive model, so click the circle next to the Additive option.

In Table 8.2, consider how we obtained CMA = 397.589 for 2000:Q3.

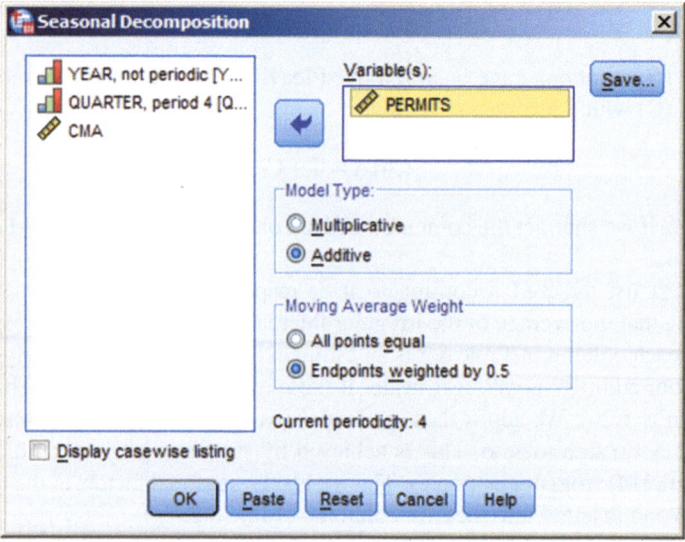

Fig. 8.5 The seasonal decomposition dialogue box

This CMA was obtained via:

$$CMA = 0.5 \left[\begin{array}{l} (2000:Q1 + 2000:Q2 + 2000:Q3 + 2000:Q4)/4 \\ + (2000:Q2 + 2000:Q3 + 2000:Q4 + 2001:Q1)/4 \end{array} \right]$$
$$= 0.125 * (2000:Q1 + 2\{2000:Q2 + \ldots + 2000:Q4\} + 2001:Q1)$$
$$= 0.125 * (2000:Q1) + 0.25 * (2000:Q2) + \ldots + 0.125 * (2001:Q1)$$

The 'weight' attached to the end points is one half that attached to the intervening readings. There is no equal weight. Therefore, in Fig. 8.5 and in the box headed 'Moving Average Weight', click the circle for the option End points weighted by 0.5.

At the bottom of this dialogue box is the Save…button, which if clicked produces the *Season: Save dialogue box* of Fig. 8.6. The method of seasonal decomposition generates some new variables, such as the seasonal factors and the seasonally adjusted data. The default option is to add them to the active data file. This would permit various plots in IBM SPSS Statistics. Click the Continue button to return to the dialogue box of Fig. 8.5 and click the OK button to run the routine.

The *IBM SPSS Statistics Viewer* will confirm the creation of any new variables and also lists the seasonal factors (or indices). Of more interest are the new variables that have been added to BUILDING PERMITS.SAV and which are shown in Fig. 8.7. The seasonal factors which were computed in the fifth column of Table 8.2 are given the IBM SPSS Statistics variable name SAF_1. The '_1' part of this variable name refers to the fact that this is the first model that we have fitted. (If we fitted some different model next during this session, the seasonal factors associated with this latter model would have the variable name SAF_2).

The seasonally adjusted data (D) in the last column of Table 8.2 are given the IBM SPSS Statistics variable name SAS_1. (There are very slight decimal rounding errors between the above output and Table 8.2, as the computations were originally to more decimal places than are reported). The SPSS variable STC_1 stands for the 'smoothed trend-cycle components' or the CMA's. IBM SPSS Statistics does not compute the CMA's exactly in the manner described above. It uses what it called *the Census Bureau 1 method* developed by the U.S. Department of Commerce. However, the end objective is the same. The IBM SPSS Statistics generated variable ERR_1

Fig. 8.6 The Season: Save dialogue box

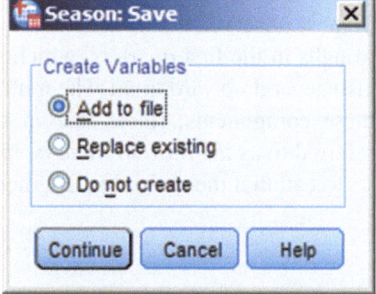

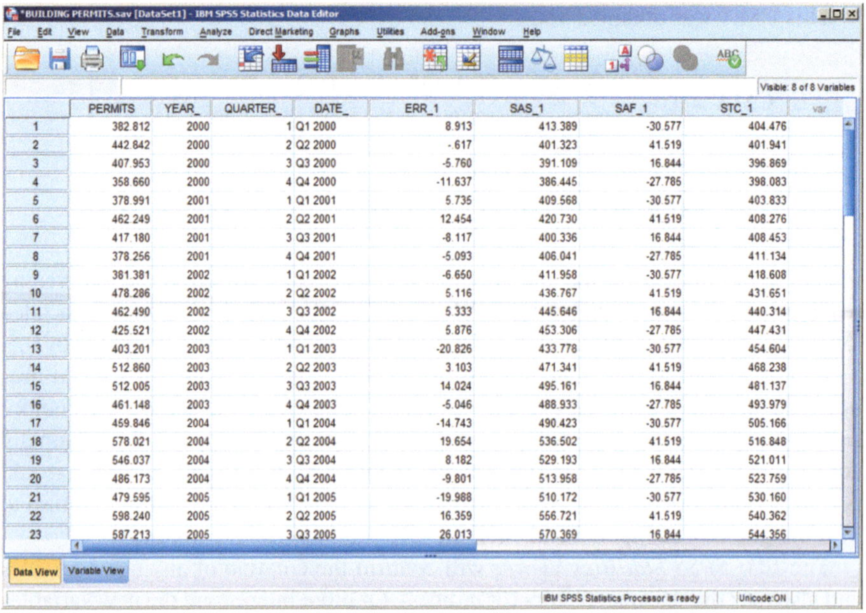

	PERMITS	YEAR_	QUARTER_	DATE_	ERR_1	SAS_1	SAF_1	STC_1	var
1	382.812	2000	1	Q1 2000	8.913	413.389	-30.577	404.476	
2	442.842	2000	2	Q2 2000	-.617	401.323	41.519	401.941	
3	407.953	2000	3	Q3 2000	-5.760	391.109	16.844	396.869	
4	358.660	2000	4	Q4 2000	-11.637	386.445	-27.785	398.083	
5	378.991	2001	1	Q1 2001	5.735	409.568	-30.577	403.833	
6	462.249	2001	2	Q2 2001	12.454	420.730	41.519	408.276	
7	417.180	2001	3	Q3 2001	-8.117	400.336	16.844	408.453	
8	378.256	2001	4	Q4 2001	-5.093	406.041	-27.785	411.134	
9	381.381	2002	1	Q1 2002	-6.650	411.958	-30.577	418.608	
10	478.286	2002	2	Q2 2002	5.116	436.767	41.519	431.651	
11	462.490	2002	3	Q3 2002	5.333	445.646	16.844	440.314	
12	425.521	2002	4	Q4 2002	5.876	453.306	-27.785	447.431	
13	403.201	2003	1	Q1 2003	-20.826	433.778	-30.577	454.604	
14	512.860	2003	2	Q2 2003	3.103	471.341	41.519	468.238	
15	512.005	2003	3	Q3 2003	14.024	495.161	16.844	481.137	
16	461.148	2003	4	Q4 2003	-5.046	488.933	-27.785	493.979	
17	459.846	2004	1	Q1 2004	-14.743	490.423	-30.577	505.166	
18	578.021	2004	2	Q2 2004	19.654	536.502	41.519	516.848	
19	546.037	2004	3	Q3 2004	8.182	529.193	16.844	521.011	
20	486.173	2004	4	Q4 2004	-9.801	513.958	-27.785	523.759	
21	479.595	2005	1	Q1 2005	-19.988	510.172	-30.577	530.160	
22	598.240	2005	2	Q2 2005	16.359	556.721	41.519	540.362	
23	587.213	2005	3	Q3 2005	26.013	570.369	16.844	544.356	

Fig. 8.7 New Variables created by the IBM SPSS additive seasonal decomposition procedure

is the difference between SAS_1 and STC_1. These variables should now be saved if they are required for future use. They have been saved in a file called BUILDING PERMIT2.SAV. Figure 8.8 presents a plot of the seasonally adjusted permit data. Although the beginning of the millennium was generally a period of expansion in the American economy, the building industry hit a market slump in 2007 and seem to have never recovered since.

8.3 The Multiplicative Model of Seasonal Decomposition

USRETAIL.SAV data file contains United States retail sales (in millions) from the first quarter of 2007 to the final quarter of 2015. The variable USRETAIL represents these retail sales. The dates have already been defined. Figure 8.9 presents a graph of these observed retail sales. There are pronounced peaks in the fourth quarters and troughs in the first quarters, which is typical characteristic of retail sales. There is also general upward trend. The multiplicative model is used in this section to isolate these components. (Note that in some texts, the multiplicative model is often referred to as *the ratio to trend or the ratio to moving average* model).

Recall that the multiplicative model is defined as:

$$Y_t = \mathrm{T} \cdot \mathrm{S} \cdot \mathrm{C} \cdot \mathrm{I}$$

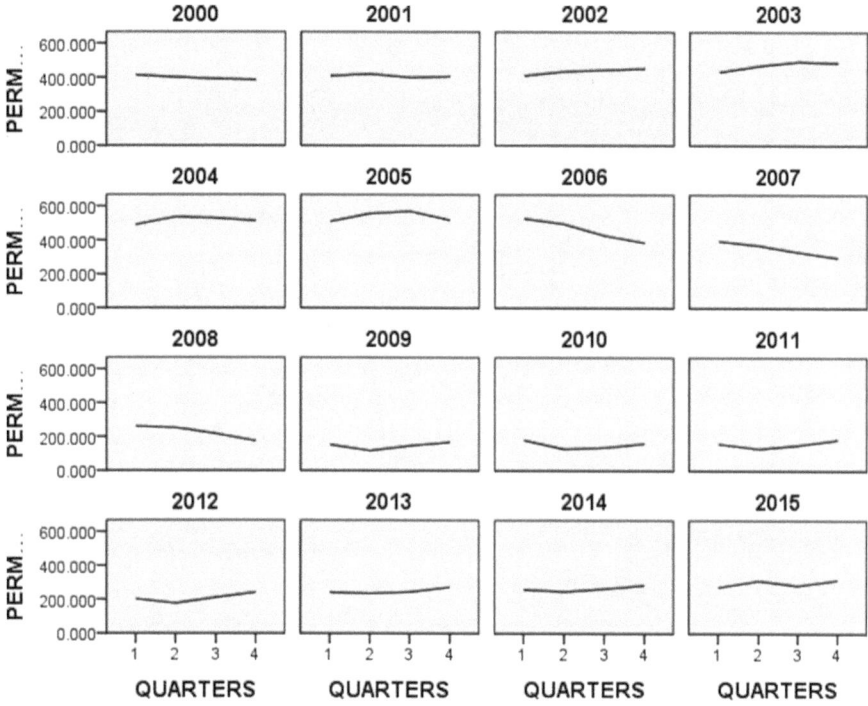

Fig. 8.8 Seasonally adjusted permit data

The stages in applying this model are similar to those employed in the application of the additive model. Four-period centred moving averages are computed in the manner described for the additive model. In the multiplicative model, we assume that:

$$CMA = T \cdot C.$$

Hence, if we divide the observed data, Y_t, by CMA then we obtain $S \cdot I$. The error component is removed from $S \cdot I$ by computing the mean for each quarter, just as for the additive model. In our multiplicative model, these averaged seasonal estimates should sum to four. If they do not, they are normalised, in this instance by multiplying each seasonal estimate by 4/(sum of averaged seasonal estimates). This generates our seasonal factors. The data are deseasonalised by dividing the Y_t by their seasonal factors. The *Seasonal Decomposition dialogue box* in IBM SPSS Statistics was selected by clicking from the Data Editor:

Analyze
   Forecasting
      Seasonal Decomposition…

Obviously USRETAIL is the variable for analysis, but this time, we are selecting a Multiplicative Model. Again, we select the option concerning weighting the end

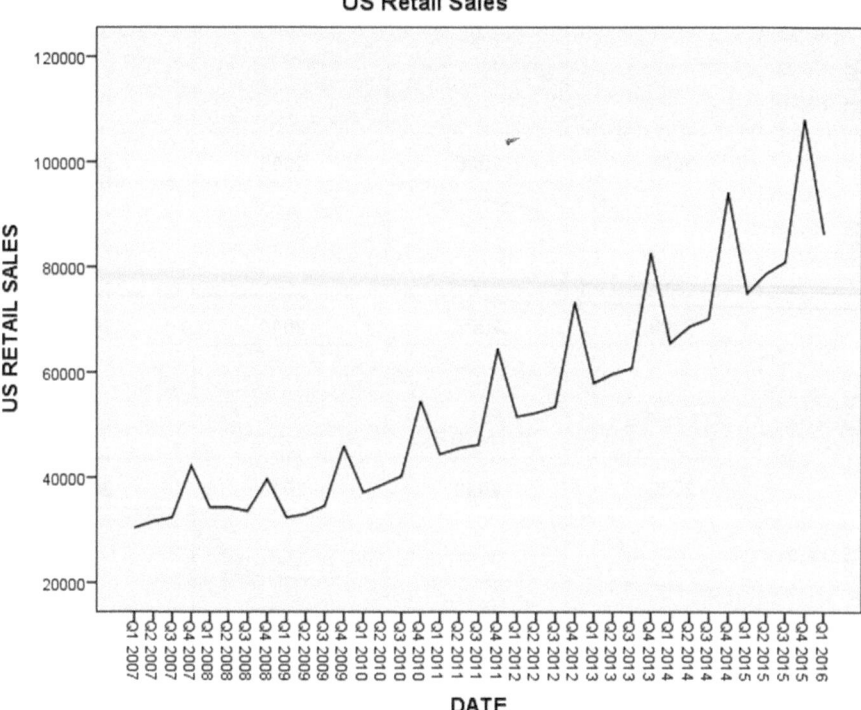

Fig. 8.9 A plot of US retail sales, 2007–2015

points by 0.5. IBM SPSS Statistics uses the same names in both the additive and multiplicative models for the new variables that may be saved. The set up for the *Seasonal Decomposition box* for the present example is shown in Fig. 8.10. For quarter 1, the IBM SPSS output shown in Fig. 8.11 reveals the seasonal factor (IBM SPSS variable SAF_1) to be S1 = 0.938. Figure 8.12 from the IBM SPSS Statistics output reports the values of seasonal factors for the USRETAIL.sav example.

8.4 Further Points About the Decomposition Method

The method is generally simple to understand and implement. It is widely used, above all for short term forecasting. Its major disadvantage is that, unlike regression, there is no statistical theory underlying the procedure.

Obviously, the decomposition methods described have involved marked seasonal components which are usually evident from a plot of the raw data. In situations where a less subjective decision is required concerning the existence of seasonality, the Kruskal-Wallis test of Sect. 6.3 may be applied. The seasonal factors such as in Table 8.4 would be input to the test. If there is no seasonal component, then the

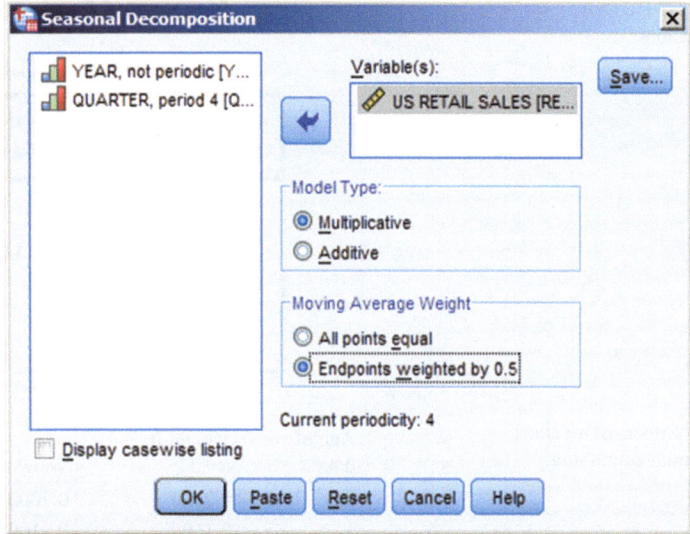

Fig. 8.10 The Seasonal decomposition dialogue box—multiplicative model

	RETAIL	YEAR_	QUARTER_	DATE_	ERR_1	SAS_1	SAF_1	STC_1	var	var
1	921266	2007	1	Q1 2007	.992	982524.210	.938	990228.669		
2	1013371	2007	2	Q2 2007	1.005	999213.213	1.014	994185.606		
3	1000151	2007	3	Q3 2007	.999	1000819.395	.999	1002099.480		
4	1060394	2007	4	Q4 2007	1.003	1011009.521	1.049	1008279.930		
5	950280	2008	1	Q1 2008	1.003	1013467.453	.938	1010132.200		
6	1028067	2008	2	Q2 2008	1.015	1013703.896	1.014	999204.559		
7	999873	2008	3	Q3 2008	1.029	1000541.209	.999	972415.211		
8	957285	2008	4	Q4 2008	.976	912702.495	1.049	934840.329		
9	828728	2009	1	Q1 2009	.973	883833.034	.938	908405.967		
10	905656	2009	2	Q2 2009	.992	893003.098	1.014	900635.414		
11	911593	2009	3	Q3 2009	1.003	912202.212	.999	909423.478		
12	966970	2009	4	Q4 2009	1.000	921936.447	1.049	922223.609		
13	876229	2010	1	Q1 2010	1.000	934492.542	.938	934098.489		
14	961240	2010	2	Q2 2010	1.001	947810.535	1.014	946676.735		
15	951077	2010	3	Q3 2010	.990	951712.599	.999	961809.724		
16	1030295	2010	4	Q4 2010	1.001	982312.286	1.049	981226.635		
17	943664	2011	1	Q1 2011	1.006	1006411.532	.938	1000368.357		
18	1034497	2011	2	Q2 2011	1.003	1020044.062	1.014	1017393.692		
19	1026969	2011	3	Q3 2011	.994	1027655.317	.999	1033834.604		
20	1097057	2011	4	Q4 2011	.997	1045965.058	1.049	1049213.298		
21	1015585	2012	1	Q1 2012	1.019	1083114.812	.938	1062621.151		
22	1078508	2012	2	Q2 2012	.994	1063440.186	1.014	1069555.867		
23	1067067	2012	3	Q3 2012	.991	1067780.114	.999	1077646.686		
24	1141069	2012	4	Q4 2012	1.000	1087927.339	1.049	1087580.660		

Fig. 8.11 Numerical output from the multiplicative model

Fig. 8.12 US RETAIL
seasonal factors

Seasonal Factors

Series Name: US RETAIL

Period	Seasonal Factor (%)
1	93.8
2	101.4
3	99.9
4	104.9

Table 8.4 Effects of α values
on exponential smoothing
weights

Age of data	Weight if α=0.2	Weight if α=0.7
0	0.2000	0.7000
1	0.1600	0.2100
2	0.1280	0.0630
3	0.1024	0.0189
4	0.0819	0.0057
5	0.0655	0.0017
Etc.	Etc.	Etc.

seasonal factors would exhibit distributions that are the same for each quarter i.e. in the ranking procedure described in Sect. 6.3, the mean ranks for each quarter would be similar. In Table 8.4, this would not be the case, confirming a marked seasonal component.

In generating forecasts for both the additive and multiplicative models of the previous two sections, we have assumed no cyclical component. It can take quite large data sets to establish the presence of this component, so some business forecasters do not incorporate it into the decomposition model. However, if the forecaster is confident that a significant cyclical component is present, it may readily be established once the trend and seasonal components have been obtained. In the additive model, T and S are subtracted from the raw data, Y_t, to obtain C+I. In the multiplicative model, are Y_t divided by the product $T \cdot S$ to obtain $C \cdot I$. The values of C+I or $C \cdot I$ are averaged to remove the error terms and only the cyclical component will remain.

8.5   The One Parameter Exponential Smoothing Model

The method of exponential smoothing is also used when short term forecasts are required, e.g. in inventory situations. The method predicts the next value of time series via a weighted combination of previous values of that series. It attaches more weight to the most recent readings than it does to the more historical data i.e. older data are less relevant and current readings are more influenced by recent past readings. There are several exponential smoothing models and this section deals with the simplest one parameter exponential smoothing model.

Denoting the gathered or observed data at time t as Y_t, the one parameter exponential smoothing model is defined as:

$$\hat{Y}_{t+1} = \hat{Y}_t + \alpha\left(Y_t - \hat{Y}_t\right)$$

And in which $\hat{Y}_{t+1}$ is the forecasted value of the time series at time $t+1$. The term α is called a smoothing constant and may take values between 0 and 1 inclusive. The name 'one parameter' exponential smoothing model derives from the fact that there is only one parameter, α, in the model above.

The quantity $Y_t - \hat{Y}_t$ is the error in forecast at time t. Therefore, under the exponential smoothing model, the error in each forecast is noted over time and a proportion α of it is added to adjust for the next forecast. The larger the error in the last forecast, the greater is the adjustment to the next forecast. The above representation of the one parameter exponential smoothing model may be readily rearranged:

$$\hat{Y}_{t+1} = \alpha\, Y_t + \left(1-\alpha\right)\hat{Y}_t \qquad (8.1)$$

both of the above equations give the forecasted value at time $t+1$. From Eq. (8.1), the forecasted value at time t may be derived, by replacing $t+1$ with t and t by $t-1$ to derive:

$$\hat{Y}_t = \alpha\, Y_{t-1} + \left(1-\alpha\right)\hat{Y}_{t-1} \qquad (8.2)$$

placing Eq. (8.2) into Eq. (8.1):

$$\hat{Y}_{t+1} = \alpha\, Y_t + \left(1-\alpha\right)\left\{\alpha\, Y_{t-1} + \left(1-\alpha\right)\hat{Y}_{t-1}\right\}$$
$$\hat{Y}_{t+1} = \alpha\, Y_t + \alpha\left(1-\alpha\right)Y_{t-1} + \alpha\left(1-\alpha\right)^2 \hat{Y}_{t-1} \qquad (8.3)$$

hence, Eq. (8.1) may be written as Eq. (8.3). Continuing this process, it is easy to show that Eq. (8.1) may be also written as:

$$\hat{Y}_{t+1} = \alpha\, Y_t + \alpha\left(1-\alpha\right)Y_{t-1} + \alpha\left(1-\alpha\right)^2 \hat{Y}_{t-2} + \alpha\left(1-\alpha\right)^3 \hat{Y}_{t-3} + \cdots \qquad (8.4)$$

The coefficients of the Y_i are all functions of α which are called *weights*. Compare the effects of two different values of α on these weights. Table 8.4 illustrates this effect, with the weights computed from Eq. (8.4). There is an exponential decline in the magnitude of the weights, which is more marked for larger values of α.

Note from any of the above equations, that if $\alpha = 1$, the most recent observation is used as the forecast of Y_{t+1}.

The numerical value selected for α is central to setting the sensitivity of the forecasts. IBM SPSS Statistics uses the gathered data and considers values of α between 0 and 1 inclusive in steps of 0.1 and 0.01 or any other incremental step selected by the researcher. This is called a *grid search*. The value of α that is associated with a minimum sum of squares of errors $[SSE = (Y-\hat{Y})]$ is considered the optimal and may be used as the basis for forecasting. Note that in IBM SPSS Statistics, the smaller the increments selected by the user, the greater the computer time spent on the task and precision is unlikely to be worth the effort.

8.5.1   One Parameter Exponential Smoothing in IBM SPSS Statistics

EMPLOY.SAV contains annual levels of employment (in thousands) in the United States from 2005 to 2015 inclusive. The one parameter exponential smoothing model will be applied to the data and a forecast generated for employment in this industry in 2016. The variable for analysis in this data file is named EMPLOY and represents the annual employment figures. The dates have already been defined.

To access the exponential smoothing procedure, click:

Analyze
   Forecasting
      Create Models…

which generates the dialogue box of Fig. 8.13. The variable under study is EMPLOY and is entered into the 'Dependent Variables' box. Under Method, choose Exponential Smoothing. Under Criteria, there are four exponential smoothing modes available. The one parameter model is referred to as the 'Simple' model in IBM SPSS and this is default as shown in Fig. 8.14 (click Continue). Clicking the Save tab in Fig. 8.13 produces the dialogue box in Fig. 8.15, save Predicted Values by ticking the box as shown in Fig. 8.15. Under Options tab, it is possible to generate forecasts by choosing the options 'First case after end of estimation period through a specified date' and by typing 2016 under Year shown in Fig. 8.16. This will generate a forecast for the employment level in 2016. Click the Continue button to return to the *dialogue box* of Fig. 8.13. Click the OK button to operationalize. Figure 8.17 presents the results of applying our exponential smoothing model in the IBM SPSS Statistics Viewer.

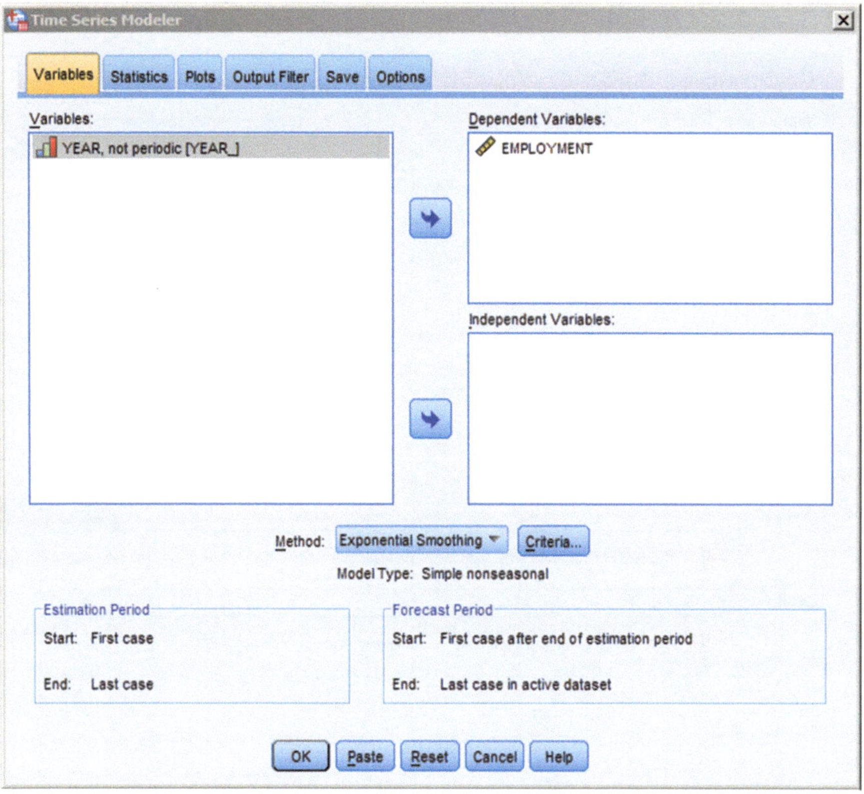

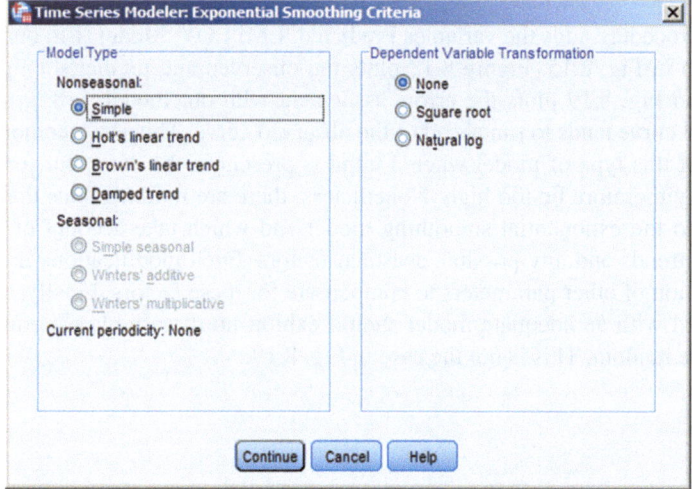

Fig. 8.13 The Exponential Smoothing dialogue box

Fig. 8.14 Simple Exponential Smoothing

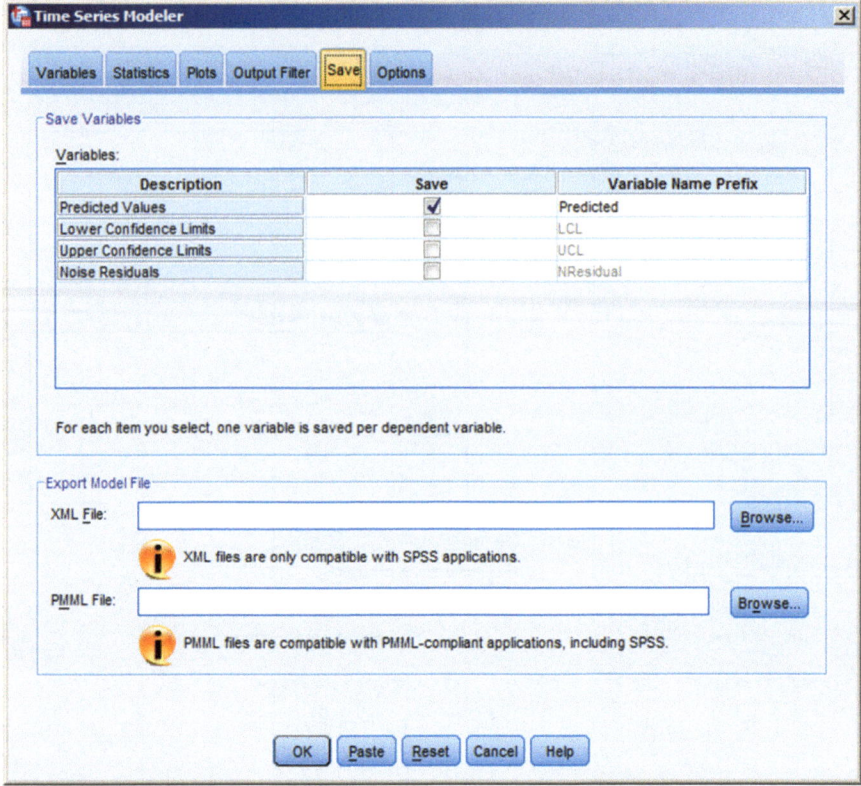

Fig. 8.15   The Exponential Smoothing: Save dialogue box

The procedure adds the variables Predicted_EMPLOY_Model_1 to our data file as shown in Fig. 8.15. Figure 8.18 plots the observed and predicted employment levels and Fig. 8.19 plots the errors associated with our model. In Fig. 8.18, the predicted curve tends to run ahead of the observed curve. This is a common characteristic of this type of model when a trend is present in the data. Our forecast for 2016 may therefore be too high. Nonetheless, there are modifications that may be applied to the exponential smoothing model and which take account of different types of trends and any possible seasonal factors. Such modifications involve the introduction of other parameters to compensate for these factors. Finally, the errors associated with an adequate model should exhibit no discernible pattern i.e. they should be random. This is not the case in Fig. 8.19.

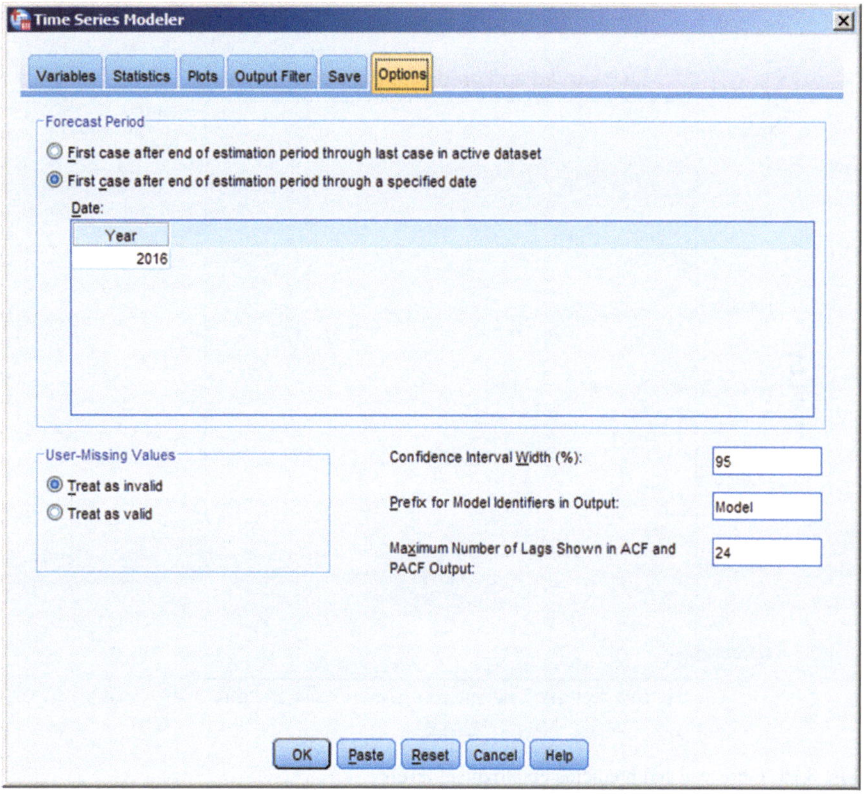

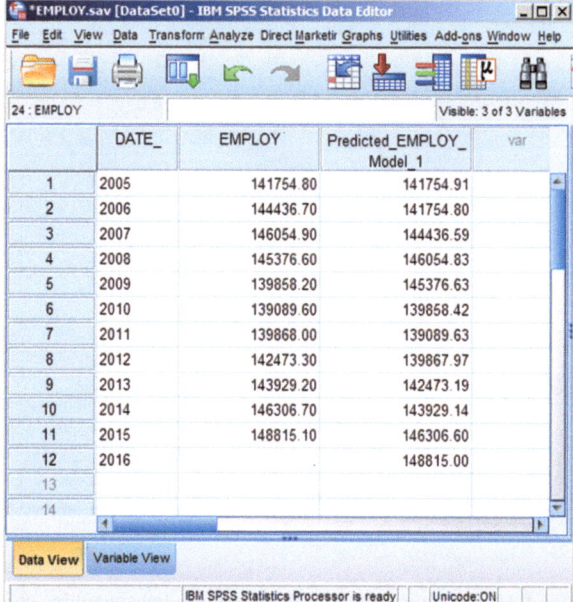

Fig. 8.16 The Exponential Smoothing: Options dialogue box

Fig. 8.17 Exponential smoothing forecast

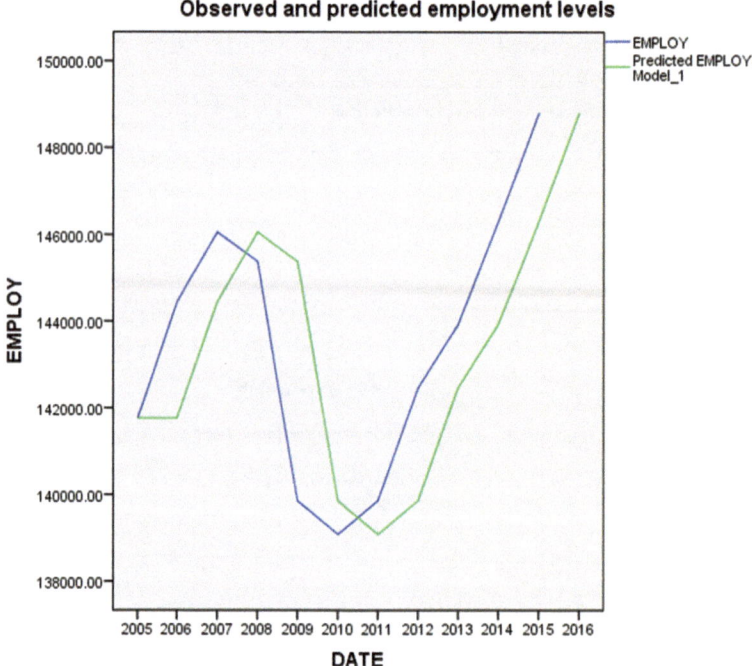

Fig. 8.18 Observed and predicted employment levels

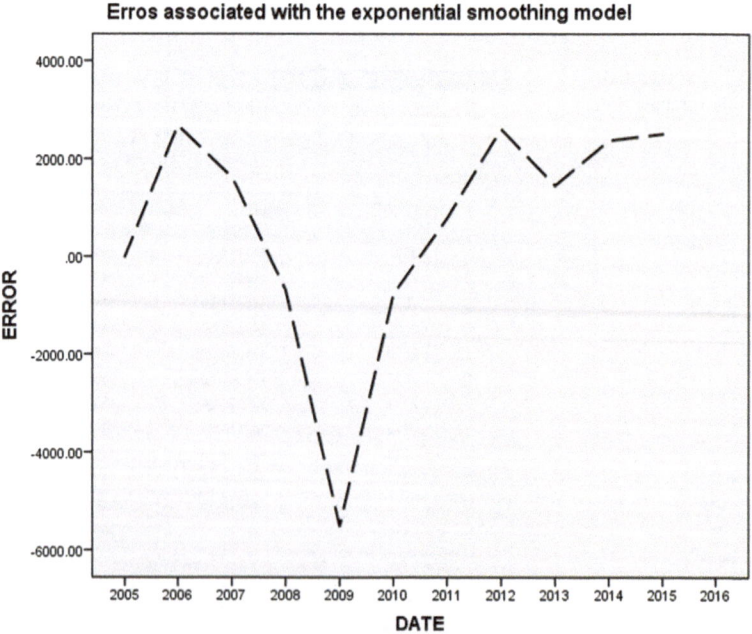

Fig. 8.19 Errors associated with the exponential smoothing model

8.5.2   Further Points About Exponential Smoothing

Exponential smoothing models are best used for short-term forecasting and they tend not to require a lot of historical data. The one parameter model is best applied to data that do not exhibit a trend that is the data fluctuate around a relatively constant value. It should be noted that the model tends to miss or lag behind or ahead of any turning points in the data. Also, as new data are gathered, it may be necessary to update α on a regular basis. As new data are produced, there are methods available for determining if a different value of α is necessary to continue the smoothing process and generate forecasts.

As with seasonal decomposition, the exponential smoothing method does not consider factors that may well influence the characteristics of the time series and thereby influence forecasts. However, the major advantage of exponential smoothing methods is their relative simplicity and low cost. For example, if a company has many inventory items and requires forecasts of each item, the fact that exponential smoothing models require only the latest smoothed figures can generate substantial cost savings when compared to competing forecasting procedures.

Part V
Other Useful Features of IBM SPSS Statistics

Chapter 9
Other Useful Features of IBM SPSS Statistics

The previous chapters have described and illustrated the operation of the IBM SPSS Statistics computer package. Elementary graphical and statistical procedures have been described. These statistical procedures are but a fraction of the methods available on IBM SPSS Statistics. For example, there are many other nonparametric hypothesis testing methods available for one, two and k independent and related samples. The IBM SPSS Statistics graphics procedures include methods for analysing industrial quality control charts. The regression methods described in this book may be extended to situations where some of the predictor variables involve nominal measurement. There are more complex time series methods other than those described in Chap. 8. In particular, the exponential smoothing model may be extended to include situations where the time series exhibits a trend or seasonality. IBM SPSS Statistics also includes a routine for applying the Box-Jenkins (ARIMA) models to time series data which will be covered in *Part II*. There are also statistical techniques available for the analysis of the structure inherent in large, multivariate data sets.

Whether using statistical routines described in this book or investigating more advanced topics such as those cited in the last paragraph, there are three features of IBM SPSS Statistics that merit comments as they will enhance the researcher's use of the package. Firstly, there is the IBM SPSS Statistics Help system. Secondly, there is the IBM SPSS Statistics journal and syntax files, by which some or all of the commands run in a particular session, may be saved and used again in later sessions. Thirdly, via IBM SPSS Statistics Preferences, the user may replace many of the default settings and replace them with user-specified ones.

© Springer International Publishing Switzerland 2016   157
A. Aljandali, *Quantitative Analysis and IBM® SPSS® Statistics*,
Statistics and Econometrics for Finance, DOI 10.1007/978-3-319-45528-0_9

9.1   The IBM SPSS Statistics Help System

IBM SPSS Statistics uses the standard Windows Help System to enable the user to understand the results obtained from analyses and to furnish information needed to operate the package. All of the dialogue boxes illustrated in this text contain a Help button, which accesses the Help system. The help offered is pertinent to the dialogue box from which the help was requested. Alternatively, the user may access Help by pressing the F1 key at any time in IBM SPSS Statistics. The window of Fig. 9.1 was generated by clicking:

Help
  Topics

from the IBM SPSS Statistics Data Editor.

Under the contents tab, select 'Help' and then, for example, 'Statistics Base Option', the user is led to the list of statistical topics presented in Fig. 9.1. The 'Linear Regression' menu provides a list of the Help topics available for the regression procedure shown in Fig. 9.2. Finally, selecting the 'Linear regression' option leads to the information presented in Fig. 9.3.

Help is available concerning the meaning of over 1000 statistical terms. After clicking:

Help
  Topics

Suppose we require help concerning the "mean". The word "mean" is typed into the box labelled 'Search', click 'Go'. This leads to a list of topics concerning the "mean" being presented as per Fig. 9.4.

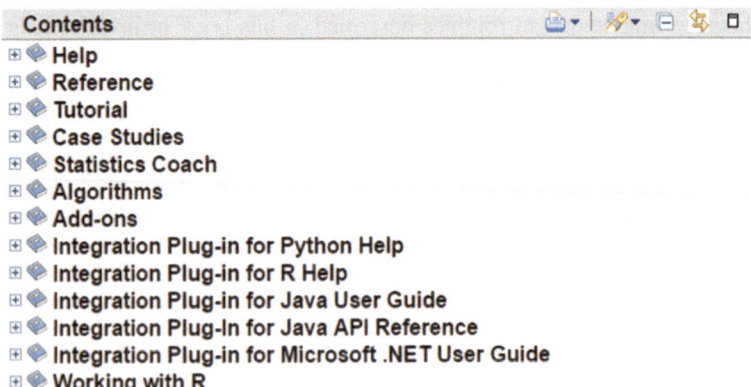

Fig. 9.1 The IBM SPSS Statistics Help topics

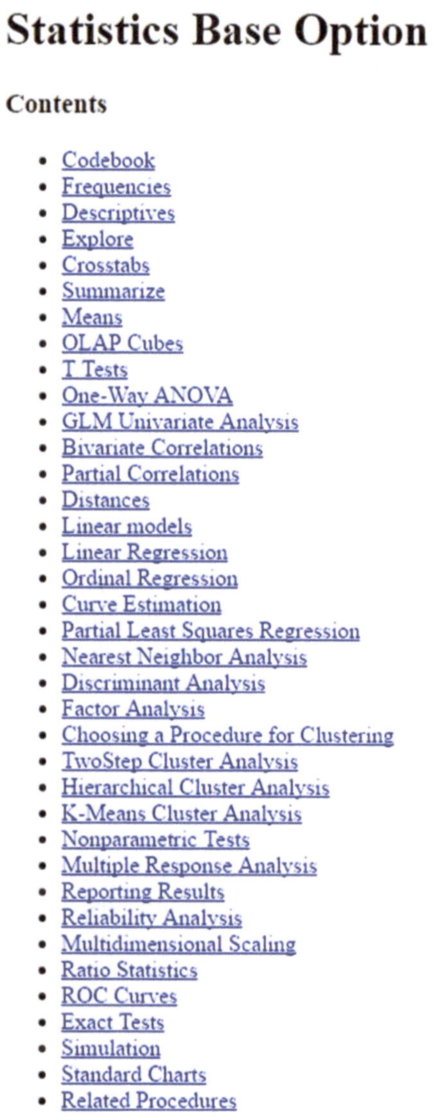

Fig. 9.2 A list of
statistical topics available
under the Help Menu

Statistics Base Option

Contents

- Codebook
- Frequencies
- Descriptives
- Explore
- Crosstabs
- Summarize
- Means
- OLAP Cubes
- T Tests
- One-Way ANOVA
- GLM Univariate Analysis
- Bivariate Correlations
- Partial Correlations
- Distances
- Linear models
- Linear Regression
- Ordinal Regression
- Curve Estimation
- Partial Least Squares Regression
- Nearest Neighbor Analysis
- Discriminant Analysis
- Factor Analysis
- Choosing a Procedure for Clustering
- TwoStep Cluster Analysis
- Hierarchical Cluster Analysis
- K-Means Cluster Analysis
- Nonparametric Tests
- Multiple Response Analysis
- Reporting Results
- Reliability Analysis
- Multidimensional Scaling
- Ratio Statistics
- ROC Curves
- Exact Tests
- Simulation
- Standard Charts
- Related Procedures

9.2   Saving IBM SPSS Statistics Syntax

A particular IBM SPSS Statistics session may involve several statistical operations being run. It is possible to save the syntax associated with a particular statistical technique for use in later sessions. IBM SPSS Statistics automatically creates syntax when the various choices are made inside dialogue boxes (Fig. 9.5).

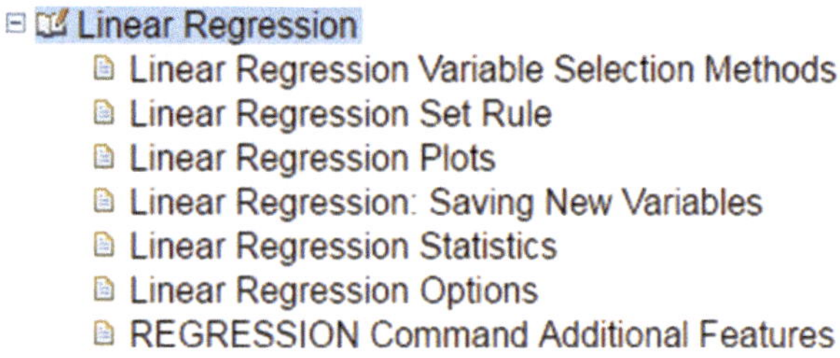

Fig. 9.3 Topics within regression under Help

⊟ ☑ Linear Regression
    ⧉ Linear Regression Variable Selection Methods
    ⧉ Linear Regression Set Rule
    ⧉ Linear Regression Plots
    ⧉ Linear Regression: Saving New Variables
    ⧉ Linear Regression Statistics
    ⧉ Linear Regression Options
    ⧉ REGRESSION Command Additional Features

It may be recalled in Chap. 7, that we regressed company returns (RETURNS) with the ratio of operating income to total assets (RATIO). We derived the regression coefficients and generated some typical diagnostic plots. Suppose that we envisage having to run this regression several times in the future. For example, we may wish to regress RETURNS against RATIO in several geographical regions to examine the constancy of the relationship. We save the syntax associated with the first regression run ad may then use it during later runs.

For convenience, Fig. 9.6 reproduces the pertinent Linear Regression dialogue box. If you click the Paste button, the dialogue box selections that have been made are pasted into IBM SPSS Statistics Syntax Editor, as shown in Fig. 9.7. If the Syntax Editor is not open, it opens automatically the first time that you paste from a dialogue box. Figure 9.7 contains the IBM SPSS Statistics syntax associated with our regression of RETURNS against RATIO and all the choices that we made. For example, a plot of the standardized residuals against the standardized predicted values, a case wise plot of all standardized residuals and a normal probability plot had been requested. Further, the unstandardized residuals (RESID) and predicted (PRED) values were to be added and saved in the data file.

To save this syntax for further use, click:

File
    Save as…

which gives rise to the Save As dialogue box for syntax of Fig. 9.8. Syntax files have the default extension .SPS in IBM SPSS Statistics, so our syntax file was named EARNINGS.SPS. This syntax file was saved on a USB on the E: drive.

In a later session, we might wish to run this regression for only those companies with, say, high sales growth. Having firstly selected these companies in IBM SPSS Statistics (via such as clicking Data then Select Cases), we need to open our syntax file. This is achieved by clicking:

File
    Open
        Syntax…

Linear Regression

Linear Regression estimates the coefficients of the linear equation, involving one or more independent variables, that best predict the value of the dependent variable. For example, you can try to predict a salesperson's total yearly sales (the dependent variable) from independent variables such as age, education, and years of experience.

Example. Is the number of games won by a basketball team in a season related to the average number of points the team scores per game? A scatterplot indicates that these variables are linearly related. The number of games won and the average number of points scored by the opponent are also linearly related. These variables have a negative relationship. As the number of games won increases, the average number of points scored by the opponent decreases. With linear regression, you can model the relationship of these variables. A good model can be used to predict how many games teams will win.

Statistics. For each variable: number of valid cases, mean, and standard deviation. For each model: regression coefficients, correlation matrix, part and partial correlations, multiple R, R^2, adjusted R^2, change in R^2, standard error of the estimate, analysis-of-variance table, predicted values, and residuals. Also, 95%-confidence intervals for each regression coefficient, variance-covariance matrix, variance inflation factor, tolerance, Durbin-Watson test, distance measures (Mahalanobis, Cook, and leverage values), DfBeta, DfFit, prediction intervals, and casewise diagnostic information. Plots: scatterplots, partial plots, histograms, and normal probability plots.

Show me

Linear Regression Data Considerations

Data. The dependent and independent variables should be quantitative. Categorical variables, such as religion, major field of study, or region of residence, need to be recoded to binary (dummy) variables or other types of contrast variables.

Assumptions. For each value of the independent variable, the distribution of the dependent variable must be normal. The variance of the distribution of the dependent variable should be constant for all values of the independent variable. The relationship between the dependent variable and each independent variable should be linear, and all observations should be independent.

To Obtain a Linear Regression Analysis

This feature requires the Statistics Base option.

1. From the menus choose:

   Analyze > Regression > Linear...

2. In the Linear Regression dialog box, select a numeric dependent variable.
3. Select one or more numeric independent variables.

Optionally, you can:

- Group independent variables into blocks and specify different entry methods for different subsets of variables.
- Choose a selection variable to limit the analysis to a subset of cases having a particular value(s) for this variable.
- Select a case identification variable for identifying points on plots.
- Select a numeric WLS Weight variable for a weighted least squares analysis.

WLS. Allows you to obtain a weighted least-squares model. Data points are weighted by the reciprocal of their variances. This means that observations with large variances have less impact on the analysis than observations associated with small variances. If the value of the weighing variable is zero, negative, or missing, the case is excluded from the analysis.

Fig. 9.4 Help for the Linear Regression procedure

Fig. 9.5 A list of help
topics related to the
statistical mean

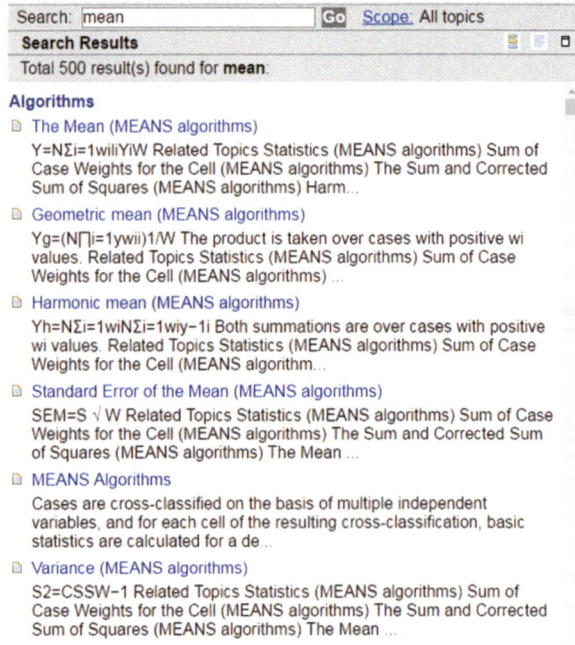

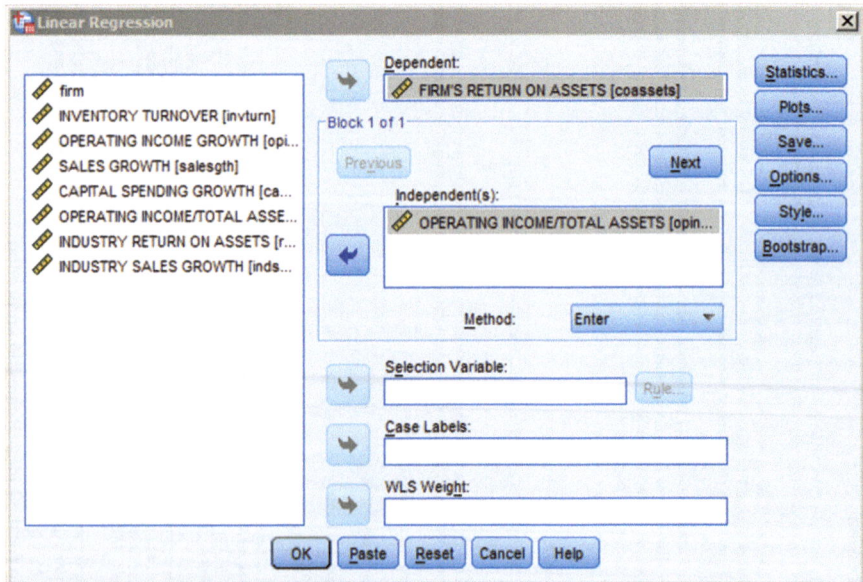

Fig. 9.6 The Linear Regression dialogue box—Returns regressed on Ratio

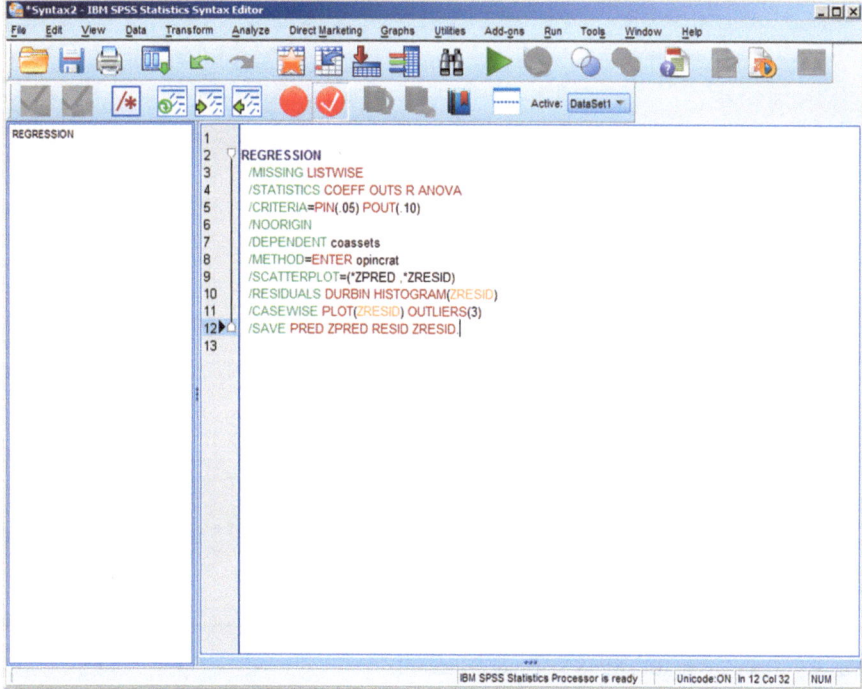

Fig. 9.7 IBM SPSS Statistics Syntax associated with the regression procedure—Returns on Ratio

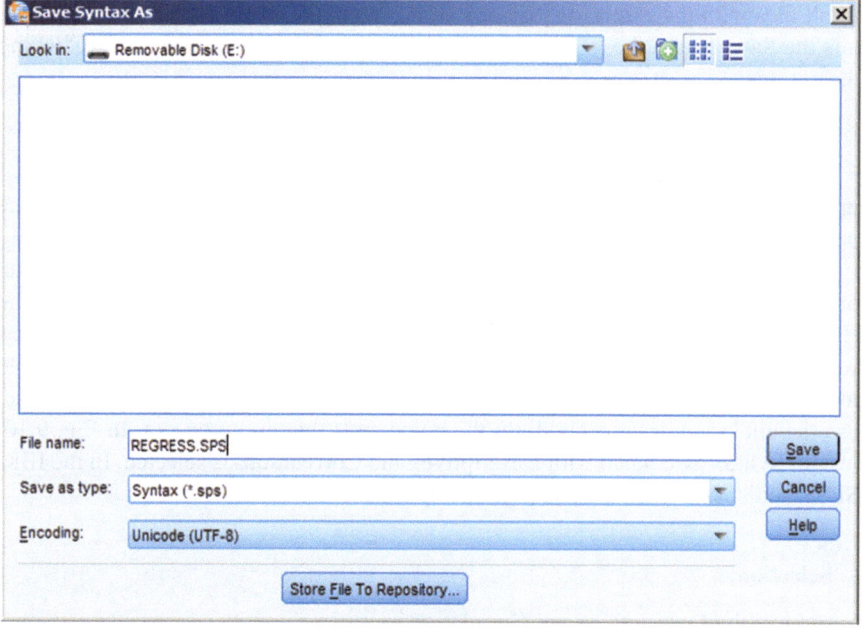

Fig. 9.8 The Save As dialogue box involving IBM SPSS Statistics Syntax

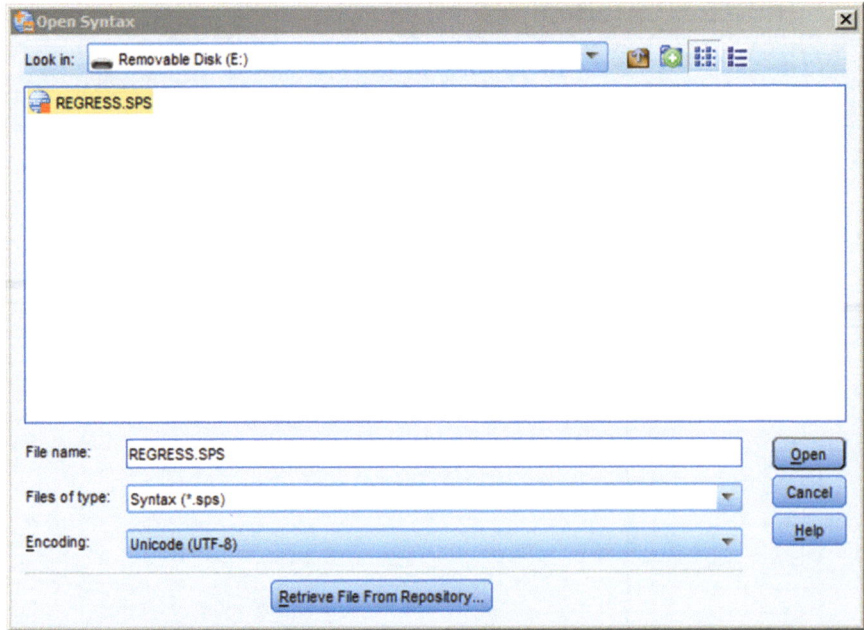

Fig. 9.9 Opening an IBM SPSS Statistics Syntax file

which produces the Open File Dialogue box of Fig. 9.9. After changing the drive to
E: all files with the extension .SPS are listed (none in this example). Once the user
hits the Save button, he is then returned to the Syntax shown in Fig. 9.7. Clicking
the options:

Run
  All…

in the IBM SPSS Statistics Syntax Editor causes the regression commands to be
operationalized.

    There will be situations where the user has saved syntax pertaining to more than
one statistical routine. For example, the Syntax window of Fig. 9.10 contains the
IBM SPSS Statistics syntax associated with the Descriptives and Correlations, as
well as the Regression routine applied to the data in EARNINGS.SAV. It is possible
to run single routines or groups of routines in a Syntax window. Using the mouse,
use the click-and-drag to highlight the commands that are to be run. In Fig. 9.10,
just the syntax associated with Descriptives and Correlations is selected. In the IBM
SPSS Statistics Syntax Editor, click the options:

Run
  Selection…

to run just the Descriptives and Correlation routines.

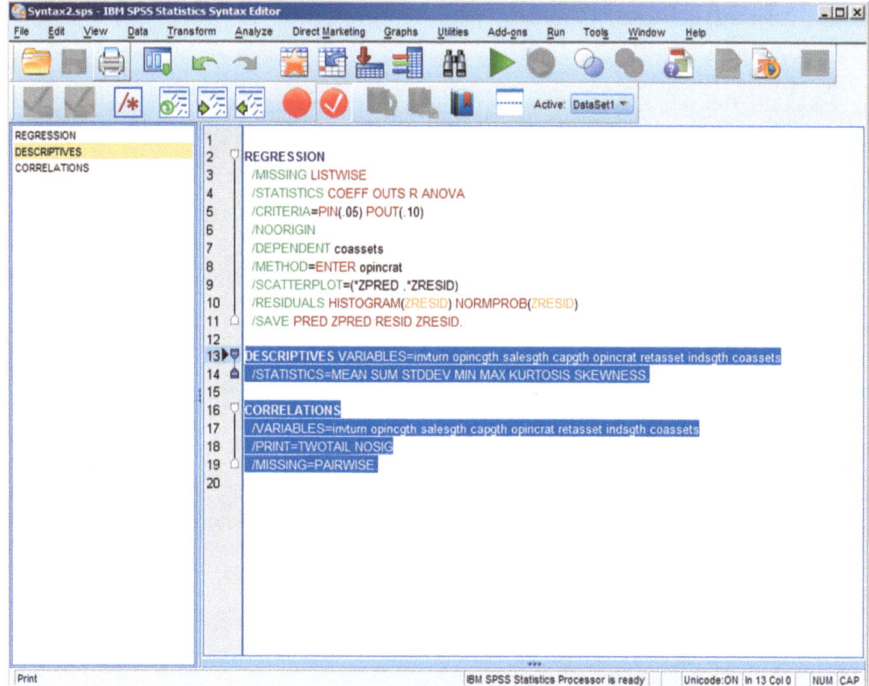

Fig. 9.10 Running part of the syntax in the IBM SPSS Statistics Syntax Editor

The method just described permits the user to select which syntax is to be saved. Syntax associated with ALL the IBM SPSS Statistics commands run during a session is automatically created and saved in what is called an IBM SPSS Statistics journal file. By default, IBM SPSS Statistics creates a journal file with the extension .JNL. It is possible to rename and re-route this journal file in the IBM SPSS Statistics Edit Options dialogue box of Fig. 9.11. This dialogue box accessed by clicking:

Edit
   Options…

and clicking the tab labelled 'File Locations'. In Fig. 9.11 and by default the directory \TEMP on the C: drive contains the journal file with the extension .JNL. This journal file is overwritten every session, as selections currently stand. To change the name of the journal file from its default of SPSS.JNL and/or to re-route it, click the Browse button of Fig. 9.11. This generates the dialogue box shown in Fig. 9.12. The file is re-routed by selecting an alternative in the box labelled in 'Save in'. Also, the name of the file has been changed to STATISTICS.JNL—a syntax file. Click the save button to return to the Edit Options dialogue box and now click the OK button to operationalize.

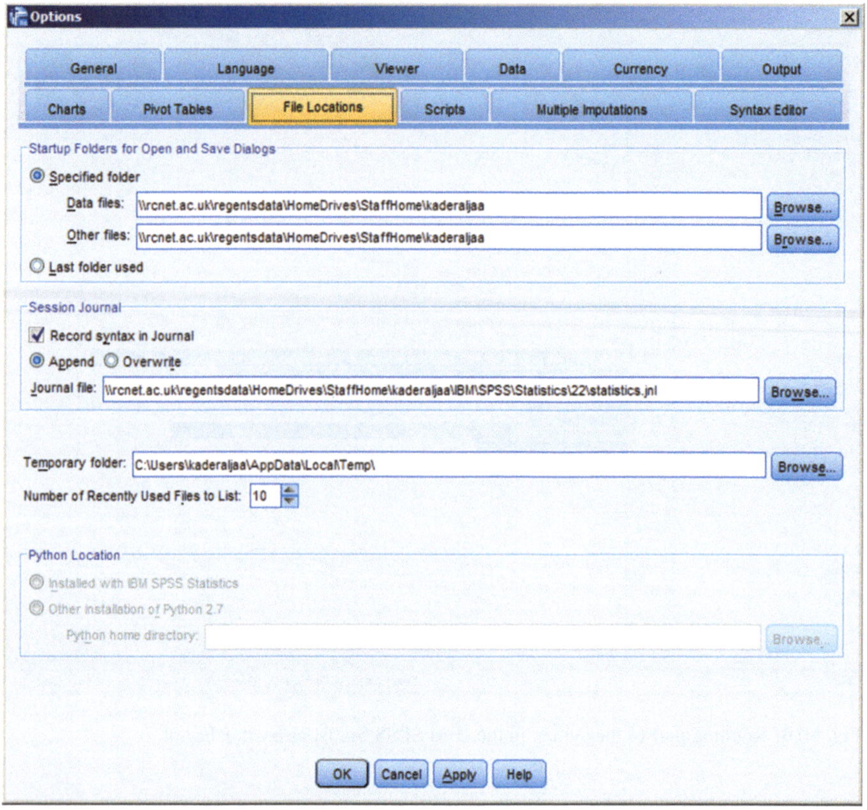

Fig. 9.11   The Edit Options dialogue box

To open STATISTICS.JNL during another session; click:

File
   Open…

and in the box labelled 'Files of type', select the option 'All files (*.*)' and click the
required name from the list of those provided. It should be noted that errors made
will be included in the IBM SPSS Statistics journal file. These are indicated by > at
the start of the pertinent line (s) of syntax and prior to running should be removed
by the user by highlighting these incorrect lines and then selecting:

Edit
   Cut

from the menu.

There are other preferences that may be selected in the Edit Options dialogue
box of Fig. 9.11 and which are largely self-explanatory. For example, if the user

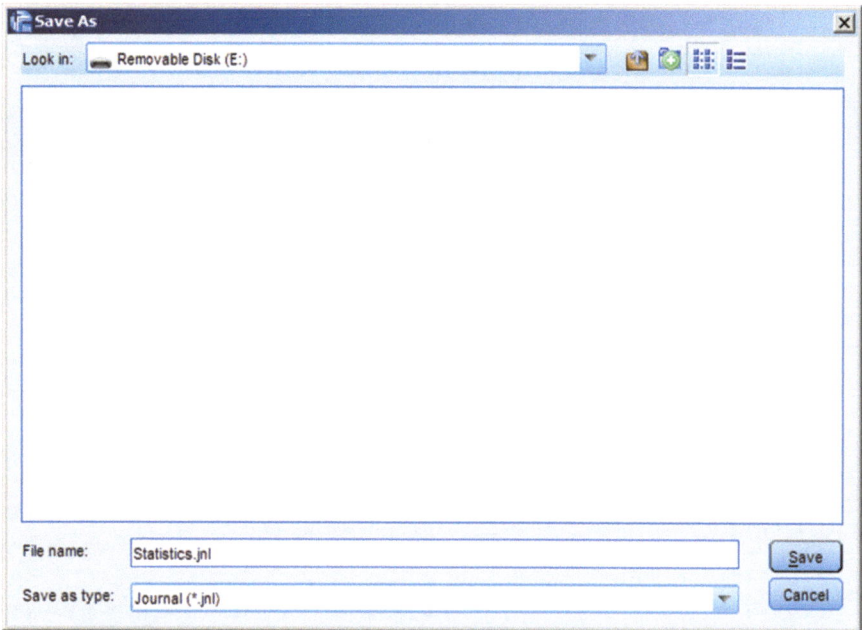

Fig. 9.12 Changing the location and/or name of the IBM SPSS Statistics journal file

clicks the 'Pivot Tables' tab in this dialogue box, it is possible to change the way in which tabulated output appears in future sessions.

In Fig. 9.13, a type of tabular output called 'Academic' is illustrated and upon clicking the OK button will become the default. Clicking other tabs in the *Edit options dialogue box* permits the user to change data display (e.g. number of decimal places), the characteristics of charts (e.g. the font employed) and the form of text output employed in the *IBM SPSS Statistics Viewer*.

9.3   The IBM SPSS Statistics Coach

For users relatively inexperienced in statistical analysis, the *IBM PSS Statistics Coach* offers some guidance as to appropriate forms of analysis in a variety of settings. The Statistics Coach asks a series of questions about the problem to be solved and the type (s) of data that have been gathered. After these questions, the Statistics Coach makes a recommendation about how to proceed. Of course, if the Statistics Coach recommends that 'Factor Analysis' is likely to be a fruitful form of analysis and the user has never heard of the method, then there must be a recourse to the relevant literature which the Statistics Coach cannot provide. As an illustration,

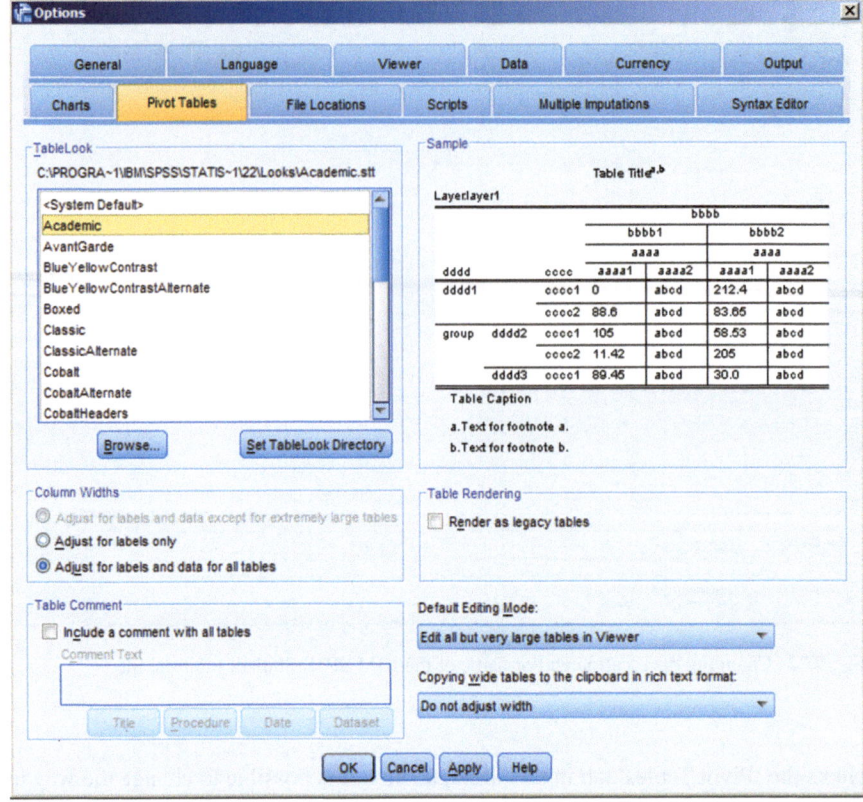

Fig. 9.13 Options associated with the 'Pivot Tables' tab

reconsider the IBM SPSS Statistics data file EARNINGS.SAV, which must be open before accessing the IBM SPSS Statistics Coach via:

Help
    Statistics Coach

which leads to the *Statistics Coach Dialogue* box of Fig. 9.14. Under the heading "What do you want to do?" suppose we wish to 'Identify significant relationships between variables'. Click this option to derive the dialogue box of Fig. 9.15.

Now we are asked "What kind of data do you have?". We have 'Scale, numeric data (interval, ratio)', so click this option to generate Fig. 9.16. Next, we are asked 'How many variables do you want to evaluate?' and select (click) 'Two (or multiple pairs of variables)' which produces the Fig. 9.17. Upon being asked "What kind of display do you want?", choose (click) 'tables and numbers' to visualise the recommendations of Fig. 9.18.

Fig. 9.14 The Statistics
Coach dialogue box

Statistics Coach >
Statistics Coach

What do you want to do?

Summarize, describe, or present data

Look at variance and distribution of data

Create OLAP report cubes

Compare groups for significant differences

Identify significant relationships between variables

Identify groups of similar cases

Identify groups of similar variables

Fig. 9.15 Further
questions asked by
Statistics Coach

Statistics Coach > Statistics Coach
Identify significant relationships between variables

What kind of data do you have?

Data in categories (nominal, ordinal)

Ordinal, rank-order, or non-normal scale data

Scale, numeric data (interval, ratio)

Ordinal dependent and scale or categorical independent variables

Statistics Coach > Statistics Coach > Identify significant relationships between variables
Scale, numeric data (interval, ratio)

How many variables do you want to evaluate?

Two (or multiple pairs of variables)

Two, controlling for the effects of one or more additional variables

Exactly three

One dependent variable and two or more independent (predictor) variables

Fig. 9.16 Even more questions asked by the Statistics Coach

Fig. 9.17 Yet even more
questions asked by the
Statistics Coach

Two (or multiple pairs of variables)

What kind of display do you want?

Tables and numbers

Charts and graphs

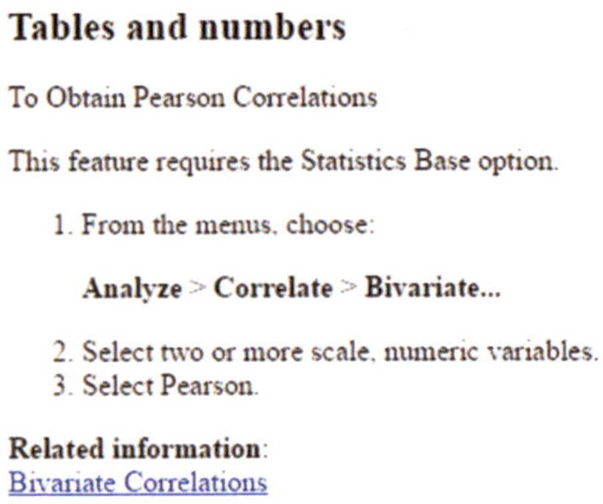

Fig. 9.18 Recommendations
made by the Statistics Coach

Tables and numbers

To Obtain Pearson Correlations

This feature requires the Statistics Base option.

1. From the menus, choose:

   Analyze > **Correlate** > **Bivariate...**

2. Select two or more scale, numeric variables.
3. Select Pearson.

Related information:
Bivariate Correlations

In answer to all the above questions, bivariate correlation is suggested for data at this level of measurement.

Possibly, the *IBM SPSS Statistics Coach* will be useful to some and indeed it may direct some researchers to appropriate methods. However, one cannot help thinking that it is yet another Windows gimmick that cannot obviate for lack of knowledge on the part of the user.

Chapter 10
Secondary Sources of Data for Business, Finance and Marketing Students

The purpose of this chapter is to describe and locate sources of external secondary data that may be of use to students of Finance, Economics, Marketing and general Business. By definition, the discussion cannot be exhaustive.

Primary data relate to information collected by the researcher himself/herself. Such data may be collected via questionnaires administered by a market research organization, for example. On the other hand, *secondary data* are provided by other agencies. The researcher is not directly responsible for collecting the information. In this latter context, *data* usually means computer readable data, since data stored in this form is more easily made available for additional research and more easily interrogated. Examples include censuses and large surveys carried out by governments and administrative data. Much secondary data is available as a hardcopy, stored by libraries, individual firms, trade associations, trade unions and the statistical arm of governments. Increasingly, more recent secondary data and related information are available on websites updated monthly, quarterly or annually.

The principal advantages of secondary data are time and cost savings. It is generally less expensive than primary research, since there is no need for the use of expensive, specialized and highly trained personnel. Secondary research expenses are incurred by the originator of the information. However, note that secondary information pertinent to a research topic may not be available (e.g. due to confidentiality) or in insufficient quantities. Some secondary data may be inaccurate, even in government publications. Also, data may be in different units than those required. For example, much data are available in index number form, so the original raw data are lost (unless you know the raw value for the base year).

Secondary data can be subdivided in terms of its source—either *internal* or *external*. Internal or in-house data is acquired within the organization where the research is being carried out. External secondary data is obtained from outside sources such as national governments.

© Springer International Publishing Switzerland 2016

A. Aljandali, *Quantitative Analysis and IBM® SPSS® Statistics*,
Statistics and Econometrics for Finance, DOI 10.1007/978-3-319-45528-0_10

Internal Secondary Data Sources

This form of data is usually an inexpensive information source. For example, internal sales and pricing data can be used as a research source. Accounting and financial records offer sources of internal secondary data. They can be invaluable in the identification, clarification and prediction of certain problems. Sales and marketing reports can include information on sales territories, locations of end-users, methods of payment, and types of product/service purchased etc. The use of such data is often used to define the competitive position of the firm, to evaluate marketing strategies or to form a better understanding of a firm's customers.

External Secondary Data Sources

There is a wealth of statistical and research data available today. Some sources are:

- National governments
- Local authorities/state governments
- Statistics agencies (like *Eurostat*)
- Trade associations
- General business publications
- Magazine and newspaper articles
- Annual reports
- Academic publications
- Library sources
- Computerized bibliographies

Most reference librarians are knowledgeable about what data are available or where to look for them. Also, government libraries and individual government departments can offer assistance.

10.1   Business and Finance Data Sources

10.1.1   Eurostat

Eurostat (http://ec.europa.eu/eurostat) is the statistical arm of the European Union. It provides a high quality statistical information service. It cooperates closely with international organizations like the UN and OECD (see below), as well as with countries outside the EU.

Eurostat contains a series of data sets arranged by "Themes" that include regional statistics, economy and finance, external trade, transport, energy and technology etc. Under these "themes" are a series of indicators:

- *Structural indicators*—employment, economic reform, general economic background
- *Euro indicators*—balance of payments, business/consumer surveys, consumer prices, trade, labour market, national accounts

- *Long-term indicators*—population data, economic and financial data, industry, trade and services, agriculture and fisheries, transport and trade.
- *Sustainable development indicators*—poverty, climate change and energy, production and consumption patterns, management of natural resources.

Of particular interest might be consumer prices (the harmonised index of consumer prices, HICP). HICP data cover all items, but are also reported for individual products and services like food, clothing, housing transport etc. Data are available monthly and 2005 is the base year.

10.1.2 OECD

OECD is the Organization for Economic Cooperation and Development (http://www.oecd.org or enter OECD as a Google search). The OECD groups 34 member countries sharing a commitment to democratic government and the market economy. It is best known for its economic, social and financial publications and its statistical data. The member countries are not all European. Data are also available for Australia, Korea, the USA, Canada and Mexico.

The OECD statistics portal provides free access to some OECD databases. There are more than 60 topics represented in the OECD statistical tables. These include economic projections, prices and purchasing power parities, finance, national accounts, transport, energy, taxation and social/welfare statistics.

The annual publication 'OECD Factbook: Economic, Environmental and Social Statistics' is available on-line.

10.1.3 UK Office for National Statistics (ONS)

The ONS (http://www.statistics.gov.uk/) provides data relating to Britain's economy, population and society at national and local levels. Summaries and detailed data releases are published free of charge. The ONS 'Key Statistics' cover GDP, retail sales, the public sector, prices and inflation, population and employment. For example, under *Prices and Inflation*, you will find the consumer price index (2015 = 100), retail price index (1987 = 100), the retail price index excluding mortgages and the producer price index. Key important publications which are available electronically include:

- *Monthly Digest of Statistics*—containing monthly and quarterly business, economic and social data. There are 20 chapters of tables covering such as the national accounts, the labour market, production (output and costs), energy, UK balance of payments, government finance, prices and wages, leisure and the weather.
- *Economic Trends*—a monthly compendium of statistics and articles on the UK economy, including some regional and international statistics. Among the areas

covered are prices, the labour market, output and demand indicators, GDP, consumer and wholesale price indices, visible and invisible trade balance, earnings and regional indicators.

- *Annual Abstract of Statistics*—a comprehensive collection of statistics covering the nation. It contains statistics on the UK's economy, industry, society and demography in easy to read tables backed up with explanatory notes and definitions. Areas covered include, parliamentary elections, overseas aid, defence, education, crime, housing, transport and communications, government finance, external trade and investment, banking and insurance.
- *Financial Statistics*—This contains data on key financial and monetary statistics for the UK. It includes data on public sector finance, central government expenditure and revenue, money supply and credit, financial accounts, balance of payments and banks and building societies.

The easiest way to access them is by entering the name of the publication in a Google search.

10.1.4   UK Data Service

The UK data service (https://www.ukdataservice.ac.uk/) is a national data service providing access and support for an extensive range of key economic and social data, both quantitative and qualitative. The UK data service provides data covering four themes:

- *Government*—large-scale government surveys, like the General Household Survey and the Labor Force Survey.
- *International*—multi-nation aggregate databanks, such as the World Bank and survey data including Eurobarometers.
- *Longitudinal*—major UK surveys following individuals over time, such as the British Household Panel Survey. This follows the same representative sample of individuals over a number of years. The objective is to further our understanding of social and economic change at the individual and household levels. The data are released via the UK Data Archive at the University of Essex and covers a multitude of topics including unemployment, household income, trade union membership, levels of state benefits claimed etc.
- *Qualitative*—a range of multimedia qualitative data sources.

10.1.5   The International Monetary Fund

The IMF (http://www.imf.org/) was established in 1946 and is an organization of 189 countries working to foster global monetary cooperation, secure financial stability, facilitate international trade, promote high employment and economic growth and reduce poverty. Data can be obtained by country. In the case of the UK, there are links to the Bank of England, the Treasury and the Financial Services Authority. It is possible to contact the IMF about the UK and you may receive free e-mails when the IMF posts new items of interest to you.

The *International Financial Statistics* section of the IMF database contains about 32,000 time series covering more than 200 countries and covers such as exchange rates, fund accounts and the main global and country economic indicators. Also available are figures relating to national debt, regional tables of balance of payments, commodity prices, total IMF resources and financial soundness indicators.

Also available are a host of reports and newsletters relating to pensions, balance of payments, imports and exports, debt etc.

10.1.6   The World Bank

The World Bank (http://www.worldbank.org) is the world's largest source of development assistance, providing nearly $16 billion in loans annually to its client countries. The World Bank produces a publication 'World Development Indicators' which is an annual compilation of data about economic development. In 2006, more than 900 indicators were produced in over 80 tables organized into six sections:

- World view
- People
- Environment
- Economy
- States and markets
- Global links

In particular, 'Global Development Finance—Summary and Country tables' is a repository for statistics on the external debt of developing countries on a loan-by-loan basis. This website presents reported or estimated data on total external debt for the 138 low- and middle-income countries that report to the World Bank's Debtor Reporting System (DRS). Also, 'Global Economic Prospects' involves annual reports with global economic forecasts and topical chapters.

10.1.7   International Business Resources on the Internet

GlobalEdge™ (http://globaledge.msu.edu/) provides current information on the business climate, news, economic landscape and relevant statistical data for 197 countries. It possesses powerful features such as the ability to compare countries using multiple statistical indicators and to rank countries based on a selected statistical indicator. There is a rich collection of country-specific international business links. The statistical data sources are found under the heading 'Research'.

Besides news items at the global and regional scales, data are available on trade (law, shows and events, company directories, logistics), money (stock exchanges, banks and finance) and current topics in international banking (globalization, outsourcing, corporate governance, risk management). The site also offers direct access to:

- *BP Statistical Review of World Energy*—statistical data on energy trends worldwide, including consumption, production and prices
- *CIA: World Factbook*—contains statistics under the headings, geography, people, economy, transport, communications and defence
- *Doing Business Surveys*—which compares the economies of 155 countries based on a wide variety of statistics, from credit availability to cross-border trading
- *FAOSTAT*—food and agriculture related statistics from the UN agency
- *Graydon International*—owned by three of Europe's leading credit organizations, it provides credit reports
- *IISI World Steel in Figures*—a publication of the International iron and Steel Institute. The site has statistical data showing current and historical trends in world steel production and consumption
- *International Trade Data Network*—data on imports and exports. It also contains information for small businesses and industry-specific news
- *OFFSTAT*—official statistics on the web. The web site links resources that provide general and country-specific statistical information. The data come mainly from statistical offices, central banks or government departments
- *Resources for Economists on the Internet*—a comprehensive document that lists all starting points for economic data research.
- *UNESCO: Institute for Statistics*—each year UNESCO collects data on education, science, culture and communication in its Statistical Yearbook. It provides access to world education indicators which allows data to be accessed by year, region or country
- *UNIDO: Industrial production Statistics*—is a geographical reference information guide from the United Nations Industrial Development Organization. The data encompass various industries and value added components.

10.1.8   Miscellaneous Sources

(a) *Stock Exchanges, option and futures exchanges and regulators*

The New York Stock Exchange (http://www.nyse.com)
The London Stock Exchange (http://www.londonstockexchange.com)
The Tokyo Stock Exchange (http://www.jpx.co.jp/)

The Chicago Board Options Exchange (http://www.cboe.com)
The Chicago Board of Trade (http://www.cbot.com)
The London International Financial Futures and Options Exchange (http://www.liffe.com)
The Securities and Exchange Commission (http://www.sec.gov)

(b) *Central Banks*

European Central Bank (http://www.ecb.int)
Bank of Japan (http://www.boj.or.jp)
Federal Reserve Bank of New York (http://www.ny.frb.org)
Bank of Russia (http://www.cbr.ru)
People's Bank of China (http://www.pbc.gov.cn)
The Federal Reserve (http://www.federalreserve.gov)
The Federal Reserve Bank of New York (http://www.ny.frb.org)
Bank of England (http://www.boe.com)

(c) *Other business/finance/economic websites sites*

Finance and Development research program—research output and links on various financial issues, especially related to developing countries (http://www.devinit.org/)
WebEc—a library containing a financial economics section, plus a mathematical and quantitative methods section (http://www.ariadne.ac.uk)
Bureau of Economic Analysis—includes a large quantity of American data, especially macroeconomic, with a historical database (http://www.bea.gov/)

10.2 Marketing Data Sources

10.2.1 Marketing UK

MarketingUK (http://marketinguk.co.uk) is a business portal for Marketing. It offers key Marketing links to Associations and Supplier Directories (e.g. the Institute of Practitioners in Advertising, the British Market Research Association), Information Theory and Education (e.g. the Chartered Institute of Marketing, the Market Research Society), and Publications (e.g. Marketing Week, Marketing, Marketing Direct, Precision Marketing).

Under the link Market Research, the user can access sites containing data, some of which has to be paid for. For example, there is access to Ipsos MORI (Market and Opinion Research International), where the user may browse the research specialisms, research techniques and publications offered by the company.

Via the Nuasoft Web Services option the user accesses Comscore Media Metrix which analyses the demographics and geography of Internet users over a wide variety of facilities. Also accessible via this option is Nielsen Online (analysis of internet marketing, audience profiling, customer analytics) and the UK Statistics Authority (free access to economic and socio-economic time series data, much of which is available monthly, quarterly and annually).

10.2.2   Datamonitor

Datamonitor (http://www.datamonitor.com/) offers an extensive series of marketing research reports. Reports are available for consumer goods, heavy industry, marketing and market research, public sector, technology and media, plus a host of other sectors. Subheadings exist for each sector. For example, the 'Advertising and marketing' section is subdivided so that the user can obtain reports concerning branding, communications, direct marketing, Email marketing. Internet advertising etc.

Datamonitor also generates country reports, forecasts and risk assessments that provide strategic insight into the geographical, political and business environments and their effects on economic performance and potential. A region such as Africa, Europe, South America is selected and then an appropriate country from it.

10.2.3   The Market Research Society (MRS)

The MRS website (https://www.mrs.org.uk/) presents the 'Geodemographics Knowledge base', which is a comprehensive directory of websites for researchers who want to use market, social and opinion research and business intelligence. Some of the links are to commercial sites but others offer free data relating to the market research process, especially the Statistical Offices of the European countries that are presented. UK Census data are available, as well as discussion papers pertaining to the 2011 Census e.g. issues related to gathering information related to religion and ethnicity.

There is a section about *real time geodemographics* which is a new subject that can be defined as the study of people according to their spatial location over time. For example, a major application of real time geodemographics is the setting of motor vehicle insurance premiums on a 'pay as you drive basis' by Aviva.

Other links that may be of interest to Marketing researchers are:

- *The Data Depot* provides a complete list of UK and European demographic, marketing and mapping data. There is a small selection of free datasets and reports.
- *The Data Archive* contains the UK's largest collection of machine readable data relating to the social sciences and humanities. Included among the huge range of datasets is the Family Expenditure Survey.

- *Experian's Business Strategies Division* provides a detailed analysis of consumers, markets and economies in the UK and around the world. Its focuses are on consumer profiling and market segmentation, retail property analysis, economic forecasting and public policy research. The division is responsible for the creation of the *MOSAIC* household-level geodemographic classification that is available in 25 countries and classifies more than a million consumers worldwide.

References

Argyrous, G. (2011). *Statistics for research: With a guide to SPSS* (3rd ed.). Los Angeles: Sage.

Brace, N., Kemp, R., & Sneglar, R. (2012). *SPSS for psychologists* (5th ed.). Basingstoke: Palgrave Macmillan.

Bryman, A., & Cramer, D. (2011). *Quantitative data analysis with IBM SPSS 17, 18 and 19: A guide for social scientist*. London: Routledge.

Burns, R. B., & Burns, R. A. (2008). *Business research methods and statistics using SPSS*. Los Angeles and London: Sage.

Coakes, S. J., & Ong, C. (2011). *SPSS version 18.0 for Windows: Analysis without anguish*. Milton, QLD: John Wiley.

Coshall, J. T. (2008). *SPSS for Windows, a user's guide: Volume 1*. London: London Metropolitan University (unpublished manuscript).

Field, A. (2013). *Discovering statistics using IBM SPSS statistics and sex, drugs and rock 'n' roll* (4th ed.). London: Sage.

Howitt, D., & Cramer, D. (2005). *Introduction to SPSS in psychology: With supplements for releases 10, 11, 12 and 13* (3rd ed.). Harlow: Prentice Hall.

Janssens, W., Wijnen, K., De Pelsmacker, P., & Van Kenhove, P. (2008). *Marketing research with SPSS* (1st ed.). Harlow: Prentice Hall.

Kinnear, P. R., & Gray, C. D. (2010). *PASW statistics 17 made simple (replaces SPSS statistics 17)*. Hove: Psychology Press.

Pallant, J. (2010). *SPSS survival manual: A step by step guide to data analysis using SPSS* (4th ed.). Maidenhead: Open University Press.

Salkind, N. J. (2014). *Statistics for people who (think they) hate statistics* (5th ed.). Los Angeles: Sage.

SPSS. (2007). *SPSS statistics base 17.0 user's guide*. Chicago: SPSS.

© Springer International Publishing Switzerland 2016

A. Aljandali, *Quantitative Analysis and IBM® SPSS® Statistics*,

Statistics and Econometrics for Finance, DOI 10.1007/978-3-319-45528-0

Index

© Springer International Publishing Switzerland 2016
A. Aljandali, *Quantitative Analysis and IBM® SPSS® Statistics*,
Statistics and Econometrics for Finance, DOI 10.1007/978-3-319-45528-0